eBusiness Essentials

About the Wiley – BT Series

The titles in the Wiley-BT Series are designed to provide clear, practical analysis of voice, image and data transmission technologies and systems, for telecommunications engineers working in the industry. New and forthcoming works in the series also cover the Internet, software systems, engineering and design.

Other titles in the Wiley-BT Series:

Software Engineering Explained
Mark Norris, Peter Rigby, 1992, 210pp, 0-471-92950-6

Teleworking Explained
Mike Gray, Noel Hodson and Gil Gordon, 1993, 310pp, 0-471-93975-7

The Healthy Software Project
Mark Norris, Peter Rigby and Malcolm Payne, 1993, 198pp, 0-471-94042-9

High Capacity Optical Transmission Explained
Dave Spirit and Mike O'Mahony, 1995, 268pp, 0-471-95117-X

Exploiting the Internet
Andy Frost and Mark Norris, 1997, 200pp, 0-471-97113-8

Media Engineering
Steve West and Mark Norris, 1997, 250pp, 0-471-97287-8

ISDN Explained, Third Edition
John Griffiths, 1998, 306pp, 0-471-97905-8

Total Area Networking, Second Edition
John Atkins and Mark Norris, 1999, 326pp, 0-471-98464-7

Designing the Total Area Network
Mark Norris and Steve Pretty, 2000, 352pp, 0-471-85195-7

Broadband Signalling Explained
Dick Knight, 2000, 448pp, 0-471-97846-9

eBusiness Essentials
Mark Norris, Steve West and Kevin Gaughan, 2000, 320pp, 0-471-85203-1

Communications Technology Explained
Mark Norris, 2000, 312pp, 0-471-98625-9

eBusiness Essentials

TECHNOLOGY AND NETWORK FOR THE ELECTRONIC MARKETPLACE

Mark Norris
Norwest Communications, UK

Steve West
Kevin Gaughan
BT, UK

JOHN WILEY & SONS, LTD
Chichester • New York • Weinheim • Brisbane • Singapore • Toronto

Baffins Lane, Chichester,
West Sussex, PO 19 1UD, England

National 01243 779777
International (+ 44) 1243 779777
e-mail (for orders and customer service enquiries): cs-books@wiley.co.uk
Visit our Home Page on http://www.wiley.co.uk or http://www/wiley.com

Reprinted May 2000

Other Wiley Editorial Offices

John Wiley & Sons, Inc., 605 Third Avenue,
New York, NY 10158-0012, USA

WILEY-VCH Verlag GmbH
Pappelallee 3, D-69469 Weinheim, Germany

Jacaranda Wiley Ltd, 33 Park Road, Milton,
Queensland 4064, Australia

John Wiley & Sons (Asia) Pte Ltd, 2 Clementi Loop #02-01,
Jin Xing Distripark, Singapore 129809

John Wiley & Sons (Canada) Ltd, 22 Worcester Road
Rexdale, Ontario, M9W 1L1, Canada

Library of Congress Cataloging-in-Publication Data
Norris, Mark.
eBusiness Essentials: technology and network requirements for the electronic marketplace / Mark Norris, Steve West, Kevin Gaughan.
p. cm.
'Wiley-BT series'—Series t.p.
Includes bibliographical references and index.
ISBN 0-471-85203-1 (alk. paper)
1. Telecommunications. 2. Electronic commerce—Equipment and supplies. 3. Electronic commerce—Handbooks, manuals, etc. I. West, S. (Steve) II. Gaughan, Kevin. III. Title.

TK5102.5.N5683 2000
658'.05—dc21 99-052541

British Library Cataloguing in Publication Data

A catalogue record for this book is available from the British Library

ISBN 0 471 85203-1

Typeset in 10/12pt Palatino by Vision Typesetting, Manchester
Printed and bound in Great Britain by Antony Rowe Ltd, Chippenham, Wiltshire
This book is printed on acid-free paper responsibly manufactured from sustainable forestry, in which at least two trees are planted for each one used for paper production.

Contents

Dedications

To Kate, Amy and Adam, who reminded me (frequently) that there is more to life than eBusiness. For now. (MTN)

To George Elrington West, 12 August 1917 to 22 September 1999. (SMW)

Preface

Electronic business over the Internet will explode in the next few years, becoming a $300–500 billion dollar market over the next few years. This is not the pious hope of a few zealous engineers—it is the judgement of just about every industry leader and market analysts.

Depending on the type of company you have, you may want to sell directly to customers or to other businesses, or you may wish to deliver access to your company's speciality information services. Regardless, the key ingredients of success in eBusiness include

- *Security*—guaranteed safe transactions and record-keeping
- *Flexibility*—the ability to extend your eBusiness solution to accommodate new products and technologies
- *Integration*—your site's tools, databases and other software and scripts keep you in touch with your customers, partners, and suppliers.

This book addresses all of these, and explains how they fit into the emerging market models that will be used to describe new ways of doing business.

This book has several unique features:

- it brings together all of the essential technology that underpins eBusiness, and places it in a useful context. The aim is to provide a 'how to' for on-line trade that is based on hard fact, sound market models and broad practical experience
- it takes the pragmatic view of a complex area that has come to be dominated by technology, not always in the user's best interests. Care has been taken here to abstract from this complexity and make the topics covered accessible and relevant to real needs. The focus is on practical application rather than technology *per se*.

It is intended for a wide range of readers:

- essential reading for those engaged in the construction, design and implementation of electronic business systems. This book provides specific technical and operational detail
- recommended for anyone who has to plan, commission or oversee an eBusiness installation. The book provides the broad understanding required to avoid expensive mistakes
- a valuable professional updating guide for system designers, integrators, technical architects, telecommunications engineers, system analysts and software designers, as well as business and information planners
- a useful text for final year and postgraduate students in computer science, electrical engineering and telecommunications courses.

It seems likely that there will be few speed limits on the information superhighways . . . and no turning back. Those who choose to stay in the laybys will be left behind very quickly. Those who choose to compete in the new age need to be aware of what lies ahead, and how they manage it. Informed choices, taken now, will pay handsome dividends as complexity and choice (inevitably) rise.

The 21st century is likely to be an 'adapt or atrophy' period for many organisations, both large and small. Those who do not embrace eBusiness may be consigned to history. It is the author's intent to inform an exciting but perilous journey.

A user's guide to this book

This book was really inspired by our wish to give a straightforward account of what has become a diverse and frequently hyped topic. We spent a long time making some sense of the various market models, standards and supporting technologies. In our quest to spare others this subtle form of torture, we have tried our best to cater for a wide range of needs, and have explained both the theory and its practical application.

Different parts of the book will, no doubt, be more or less relevant to different people. Some have been written to outline general principles, others to recount a very specific area, such as security or payments. To help you select a suitable path through the book, here is our summary of the joys that we think each chapter contains.

	Technical Content	General Interest	Specialist Detail
Chapter 1	*	****	*
Chapter 2	**	***	***
Chapter 3	***	***	****
Chapter 4	*****	***	*****

Chapter 5	*****	***	*****
Chapter 6	****	***	*****
Chapter 7	***	***	****
Chapter 8	***	****	***
Chapter 9	*****	****	**
Chapter 10	*	****	*
Appendix 1	**	*****	**
Appendix 2	***	****	**
Glossary	*	***	**

If you read from beginning to end, we trust that you will find an interesting story. If you already know a bit about eBusiness, the individual chapters are structured as reasonably self-contained explanations of a specific topic. Whichever route you take, we have included a sizeable glossary to ensure commonality of interpretation.

About the Authors

Mark Norris is an independent consultant with over 20 years experience in software development, computer networks and telecommunications systems. Over this time he has managed dozens of projects to completion, from the small to the multi-million pound, multi-site, and has worked for periods in Australia, Europe and Japan. He has published widely over the last ten years with a number of books on software engineering, computing, project and technology management, communications and network technologies. He lectures on network and computing issues, has contributed to references such as Encarta, is a visiting professor at the University of Ulster, and is a fellow of the IEE. Mark plays a mean game of squash, but tends not to mix this with business of any kind. He can be found at: mnorris@iee.org

Steve West leads the development of BT's eBusiness products, and has over 15 years experience in communications, software and information technology. He has worked on a variety of projects, including the management of BT's work within the Open Group and generation of corporate IT strategy. Steve co-authored one of the existing books in the Wiley/BT series, *Media engineering*, with Mark Norris.

Kevin Gaughan is Head of Product Development within BT's eBusiness unit. With BT for 17 years, Kevin's career has spanned the management of major technology and business programmes, the definition of BT's IT strategy, the creation of the internal eMail and intranet service for BT's 100,000+ people world-wide and most recently leading the technology focus of a major review of BT's eBusiness Strategy. Passionate about the transformation of business through IT, Kevin has a wealth of first hand experience in how both large and small enterprises can embrace and exploit technology trends.

Acknowledgements

eBusiness is a complex and dynamic area, and no-one can know everything about it. So this book draws on a number of back-room heroes for its completeness and authority, and the authors would like to acknowledge the contribution of the BT eBusiness Engineering Team. Their ideas and experience underpin much of what is presented in this book, and some of the words and wisdom of Paul Putland, Dave Johnson, Tony Fletcher and Ian Videlo can be found scattered through the pages that follow.

Thanks are due to Ann-Marie Halligan, Sarah Lock, Laura Kempster and the team at John Wiley who supported and encouraged the production of this book.

1

Electronic Trade

Hegel was right when he said that man can never learn anything from history

George Bernard Shaw

President Bill Clinton has estimated that, worldwide business worth about $375 billion will be handled electronically by the year 2002—a figure disputed by Nicholas Negraponte, who reckons that the President has underestimated it by a factor of three. Whichever opinion you go with, it is now generally accepted that a major part of the way we do business from here on will involve computers and networks.

Why should anyone be in the slightest bit interested in this? After all, people have transacted business for centuries, and we have had the wherewithal to exchange the electronic data to support it for over twenty years now. What has happened to make electronic business the intriguing, and very lucrative, proposition that it is now perceived as?

The answer can probably be condensed into a single word—Internet. Before this (by now familiar) phenomenon's dramatic growth, electronic trade was routinely conducted, but it relied on complex, expensive and proprietary equipment that ran over private data networks. There was little to commend it, and although it succeeded in making a steady income for some niche companies, it failed to take the business world by storm.

Over the last few years, the Internet has matured to provide a global infrastructure, accessible to a vast number of people. It makes global markets a reality, even for individuals and small businesses. More significantly, the Internet makes it possible to transform electronic trading from an expensive and specialised process into a cheap and realistic proposition for the masses.

However, there is more to electronic business than having a worldwide network of computers that can readily share and exchange information. There are some prerequisites to successful eBusiness. For example, secure transaction processing, the establishment of trust when personal contact is absent, the handling of intangible, information-oriented products, and the slick integration of presentation, billing and fulfilment are vital. They have to be packed into a straightforward and intuitive package if eBusiness is to fulfil even Bill Clinton's modest expectations: and that is what this book is all about.

Before we get to the exotica of catalogues, nanopayments and digital certificates, we need to set a clear baseline. So, in this chapter, we aim to explain the nature of the digital marketplace by defining what we mean by eBusiness, what forms it can take, who the players are, and what the essential elements of an electronic market stall are. A key point is that you need more than an investment programme to succeed: a shift in attitude and understanding are also required. Well-established physical assets and business processes also need to be recast into electronic services. So we will introduce the key concepts such as security, trust and encryption that are explained in some detail later on. First, though, some definitions.

1.1 WHAT IS eBUSINESS?

There is no universally accepted definition of eBusiness. In this book we assume that it embraces all aspects of buying and selling products and services over a network. The essential characteristics of eBusiness are that the dealings between two parties, be it business to consumer or business to business, are online transactions, and that the key commodity being traded is information.

In effect, we see eBusiness as the gateway to a deal—it is a transaction that may, but doesn't necessarily have to, lead to the delivery of a physical product. There are several commonly used names for eBusiness, the most popular being eCommerce and eTrade. Some of the more academic treatises attempt to distinguish the terms (for instance, eCommerce is sometimes limited to the buying and selling of goods and the flows of associated information and funds; eTrade can be viewed as covering only supplier to supplier transactions). We have used such terms as synonyms here, and we take them as referring to the same thing.

Whichever term is used (and we will go from one to another for the sake of variety), there *is* a clear differentiation between the 'e' and the 'business', 'commerce' or 'trading'. The former is a question of technical capability, the latter the way in which that capability is applied. Put them together and you end up with something more than the sum of their combined parts. New possibilities and requirements emerge. In this book we will concentrate on the technical aspects of eBusiness, but the rest cannot and will not be ignored.

Therefore, our focus as we explore the emerging world of eBusiness will be on:

- Technical aspects—the hardware, software and networks that are needed to connect a community of interest and allow them to share information. This also covers the design and presentation of that information. An important part of eBusiness technology is the specialised software used for payments (billing, charging, invoicing, account management), security (authorisation, authentication, privacy, data integrity and audit) and service support (problem management, configuration control and order handling). These will be explored in some detail.
- Business model—how businesses interwork, and how this influences the way in which they are established and the way in which technology is deployed. In this respect, we examine the typical flow of orders, fulfilment and payments, how various players co-operate to provide an end product or service to the consumer, and the various ways in which a virtual market is established.

The last part of this section will introduce the structure of the book, which starts with a description of the marketplace as a whole, before explaining what a shop within that marketplace consists of. From here we move on to explore the elements that make the shop viable—catalogues that advertise its wares, security measures that instil customer confidence, and the shop window features that attract and retain customers. Putting these various components together, we go on to build a comprehensive picture of eBusiness.

Figure 1.1 gives some idea of how the key elements fit together. It should be said that the perspective is intended to be general, and should fit business-to-business trade as well as the case where individual consumers interact with an online business. So you can 'daisy chain' the picture, such that someone who takes the role of *supplier* to one set of customers may also take the role of *customer* to a different supplier further back in the chain. In any case, they all have to be on a shared network, work from the same catalogue, have some means of delivering goods, and be able to settle up after the transaction.

What is not shown in Figure 1.1 are the intangible attributes that make a business operate (more or less) effectively. Later on we explain the essential technology that needs to be deployed to establish an eBusiness, how trust and security are established, and we illustrate some real instances through case studies.

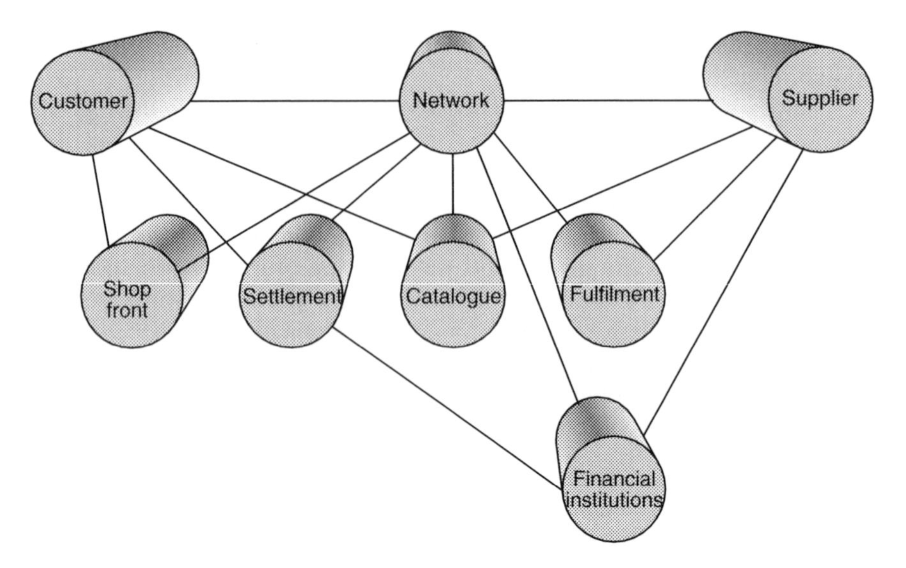

Figure 1.1 The essential elements of an electronic business

1.2 THE WHISTLE-STOP TOUR

So far, we have described eBusiness in broad terms as a mass-market capability that is enabled by a combination of the Internet's global reach and the vast resources of traditional Information Technology (IT). Given this, it should come as no surprise that it is a multi-faceted beast (or, potentially, a many-headed monster!). Many of those facets are technical, but others are not.

To get a grasp on the overall scope and nature of the eBusiness proposition, this section looks at the constituent parts of trading over a network. There is a lot more detail on each of these areas later on. For now, though, we have the surfer's guide.

The marketplace

Before we think about business, we should first think about the market where that business is conducted. So, what is an electronic market? It can be viewed as a direct parallel of the familiar shop, store or emporium. It is, in essence, a virtual trading area where deals are struck over a network. The 'shop-front' is the computer and the server is the warehouse. In fact, there is an electronic analogue of virtually all of the items you'd find in a conventional market—including bogus traders, inferior goods and dubious bargains.

The size and scope of the marketplace is, however, a little different from the familiar high-street model. In eBusiness, unlike many other areas of high technology, size really does not matter. It is quite possible to conduct a large volume of business over a wide area with little overhead. The 'information smallholder' can compete on an equal footing with the multinational corporation. In fact, it is often not that easy to distinguish between the two. People who have encountered Amazon.com on the Internet probably have no idea how it compares in size with, say, W H Smith, who have bookshops in most large towns in the UK, or Blackwells, who position themselves as global suppliers of academic books.

To be an effective player—as well as the basic technology to enter the market—you need a brand, some content, service support and a means of fulfilment (i.e. delivering the goods). This does not imply that all have to be owned. An eBusiness that looks like a cogent entity to the consumer may in reality be a host of co-operating suppliers. One supplier might provide the online content and another the application hosting (those clever facilities available on the user's PC and server). The delivery vans might be contracted from the Post Office or Federal Express, and they might deliver a set of products branded by someone completely different. In this respect, the electronic market is a more complex beast than the traditional one of manufacturer, retailer and wholesaler.

Furthermore, the nature of eBusiness lends itself to more than one trading model. For instance, the age-old ideas of a marketplace owned by one organisation, populated by (authorised) traders is quite tenable (Barclay Bank's Barclay Square and NatWest's Buckingham Gate are good examples of this). This model can be extended to add the idea of having 'guilds' that control standards within their particular area (but more of this later).

There are many other trading models, such as auction (conventional as typified by e-Bay, Dutch, sealed bid) and then there is barter—eBusiness draws on a lot of history! All have their merits and are suitable to a certain type of trading. Most online trading models have a physical dual, some don't.

Whatever the market looks like, it can often be categorised by the dominant party. In seller-driven markets, it is the large, dominant vendor who sets the price, not really for negotiation; in a buyer-driven market, there are many people selling into the market, and a small number of dominant buyers who take best bids (for example, UK supermarkets are often accused of having undue influence—downwards(!)—on the price of agricultural produce). It is these dynamics that determine the appropriate technology for eBusiness. There are also open markets in which the buyer and seller negotiate, or a free market, where the behaviour of the market itself sets prices. Insurance brokering is an example of this, as are the optimisation packages used for pricing and selling airline seat capacity.

eShops

The electronic shop can be thought of as the 'look and feel' of the screen that fronts the customer. Just as with high street stores, the aim is to entice the customer to browse and, ultimately, to buy.

Although unlikely to supplant real-world shops, the online variety can provide features that seem likely to promote their growth. As well as being readily available and easy to search, they can provide some measure of mass customisation. For instance, the made-to-measure shirtmaker, Charles Tyrwhitt has an online shop that can apply a buyer's previously entered measurements and preferences as it mails each order from the Jermyn Street shop. Every customer is thus treated as an individual, but there is capacity (at least on the customer handling side) to cater for a mass market.

The fundamental prerequistite for presenting products and services online in the eBusiness world is the catalogue. These are central, and are the electronic equivalent of a shop's shelves, goods, special offers and departments. The catalogue is the online representation of what is 'for sale' (or more correctly, what is available for trading).

It is important to appreciate that there are different scales of catalogue. They range from a set of web pages and a simple script that allows orders to be taken, through mid-range catalogue products that are characterised by a pre-defined structure of product categories and sub-categories, up to large scale corporate catalogues that are customisable. In this last case, there is usually back end integration with inventory, stock control and ordering systems.

Another important point about catalogues is that they are different for buyers and for sellers. The former is a virtual directory that allows the buyer to look at and judge a range of competing products from a number of different suppliers. The latter is a structured set of information that represents what a particular supplier has to sell. The technology used to represent these different types of catalogue has to match—that is, it either has to be optimised for one seller and multiple buyers, or *vice versa*. More on this later.

One further differentiation in catalogues that should be made is that of business-to-business as opposed to business-to-consumer. The consumer-oriented catalogues tend to be stronger on presentation, as they usually have to sell on the basis of eye-appeal. The business catalogues are more focused on quick access to another business' needs, and these tend to be high volume and fairly routine. For instance, many supermarkets issue stock replenishment orders from their automated stock control systems—they buy a lot of carrots on a regular basis!

In short, the distinction between consumer- and business-oriented catalogues is akin to food mart against delicatessen.

Payment

So to the core of any business (and eBusiness is no exception)—profit. Trade, commerce and business only exist to satisfy the needs and desires of the participants. This means one party getting something they want in exchange for something the other party wants; and usually it is money that fuels the desire to trade.

By its very nature, eBusiness needs to emulate in some way the customary direct exchange of cash for goods. Attitudes to the use of different payment mechanisms are changing (and vary when considering Europe, the US, Asia-Pacific or the whole world). A priority in establishing an eBusiness is to put in place an acceptable mechanism for payment. There are many technical options, and to choose appropriately, factors such as scale and acceptability all need to be carefully examined.

By way of illustration, there are various scales of payment to be considered: items such as books are regularly purchased electronically (e.g. through Amazon.com), and it is common for a single customer to buy a single book and pay with a single credit card transaction (probably in the $10 + range). However, the cost of processing credit card transactions is quite high, and may not be economical for the purchase of small-value items such as consumer reports or individual music tracks, which would typically be in the $1–10 range. Technologies exist for handling these small-value payments (known as 'micropayments'), and lumping them together into single credit card transactions. The range below $1 is often referred to as the area of 'nanopayments'. Such payments may represent, for example, the price of viewing an individual Internet page of information (e.g. a particular share price).

Not only are different technologies required to aggregate these various categories of small-scale payments, but also the eBusiness will need to put in place different strategies for such things as handling account queries and dealing with bad debts. For instance, a sensible way of dealing with lost nanopayments would be to write off the debts, and blacklist the offenders—the cost of recovering the debt would be too high.

Settlement

It is all very well taking a 'virtual' payment for goods or services offered over the Internet but, at some point, this must be converted into dollar bills, Euros, or some other tangible form of money. Hence, we need a gateway between the virtual world and the real world—a payments gateway. This can be effected with automated connection to Merchant acquirers (e.g. Barclays Merchant Services), the BACS system, direct debits, or other systems, such as PC-EFT, WorldPay and Clear Commerce.

Of course, traditional settlements are not mandatory. In the past, people have used all kinds of different objects to represent money: leaves, sticks, beads and even bits of metal or pieces of printed paper! In the electronic world there are additional possibilities, such as token-based systems, where you first buy a number of tokens, in your chosen currency, and are then free to spend them on goods which are priced in tokens. An example of this is web 'beanz' which are rather like on-line loyalty card points. Electronic wallets and smart cards such as Mondex provide another alternative. Some of these types of electronic cash are fully portable, whereas others retain your wallet on your PC or on a server.

Presentation

As with catalogues, the way in which online products and services are presented depends upon the market. For the electronic shop, the art of window dressing is not really one that has transferred online as yet. Already the ease of use of online systems is coming under the scrutiny of consumer watchdogs such as *Which? Magazine*, who report a significant impact (both positive and negative) of presentation quality.

When there are many suppliers (and this is worldwide), the look and feel of the shop will make a big difference, as will the ease of use. In much the same vein, business-to-business transactions will need to be reliable and easy to use, in their own way.

The design of online information is very much in its infancy, but there are some basic guidelines for getting the right presence and operation for the job in hand (and one of the best ways to see what you should do right is to look at other people's mistakes). Navigational dead-ends, inconsistent and out-of-date information, lack of an overall information map, frustrating and non-intuitive structures and poor search/browse capabilities all put customers off. The electronic window dresser is one of a number of new skills being driven by eBusiness.

One of the interesting twists to presenting what you have to offer online is that it is possible to find out exactly what each customer has looked at, what they purchased and when. This would consume a huge amount of time and video to do in a conventional shop, but consists of little more than analysing system logs for an eBusiness. Information about individual customers, their browsing and buying patterns, is an important piece of feedback on how to go about presenting your goods.

Security

In the 'real' world you go into a shop (about which you can make some judgement from its location, size, type of premises, how long it has been there, and so on), and you hand over cash in return for goods which you carry away. The risks are very small, and even if things do go wrong, you can usually exchange the faulty goods. You know where to go back to (bricks and mortar rarely move overnight) and who to talk to. If you dont get satisfaction you can even, as a last resort, make a scene: standing in the middle of a busy store, telling an assistant, as loudly as you are able, how badly you have been treated, can elicit rapid solutions to your problems.

When trading over the Internet, things are not so simple: the dream of the virtual trader can suddenly become a nightmare. For example, how do consumers know (before they hand over their credit card details) that the company they are dealing with is reputable, and is what it purports to be? Conversely, how does the trader know that the consumer is not using stolen credit card details?

Furthermore, how do both parties ensure that their transaction takes place privately without someone else snooping on it or, even worse, tinkering with the transaction details while they are in transit across the network? And when things go wrong (as they inevitably do), what sorts of mechanism are there to ensure that both parties fulfil their obligations?

Secret codes have long been used to ensure the privacy of information that must be sent by untrusted carriers (or public networks): in 1660, Samuel Pepys was certainly devising such codes for use by English aristocrats communicating with the exiled King Charles. In eBusiness, the principle is the same, although the codes (or cryptography) need to be a lot stronger because the potential eavesdroppers are sophisticated, and are assumed to have access to powerful computers and software.

There are many ways of establishing a secure communications link. Basically, for two people to be able to communicate using a code, they must 'share a secret' (the key to the code), and this secret must be denied to everyone else. Pepys must have had the same problem: how to agree with the King and his supporters in Holland what code would be used for communications from England. The same problem exists for today's cryptographers: how do a trader and a consumer (who will probably never meet) agree on a secret code that cannot be guessed by anyone else? The answer is to use a type of coding which employs a so-called 'asymmetric code'. This is one in which the message is encoded using one key and decoded using a different key. In this instance, getting a copy of the encoding key (the so-called 'public' key) does not allow you to decode the message—for that you need the other, 'private', key.

Even if a link is secure, you need to know that it is the one you wanted. So, when you access an Internet site that calls itself First National Bank, you want to be sure that it is truly what it purports to be before you engage in a financial transaction with it. The way this is done is to have the site certified by a 'trusted third party', who checks the authenticity of the site and provides certified copies of that site's 'public key' for communicating securely. There are a number of types of certificates that are issued by such trusted third parties, each with its own level of trustworthiness and area of application.

Finally, what happens if something goes wrong? There has to be some means of checking who promised what to whom, and when. The concept of non-repudiation (proving that a deal was struck) is one that needs to be transferred into the world of eBusiness. This is where verifiable transaction records and digital signatures fit.

A successful eBusiness should consider and have viable policies, approaches and solutions to all of the above issues. In addition, there is the small but important practicality that a certain amount of infrastructure has to be established before you can actually begin to trade online. As this book unfolds, we deal with each of the key aspects of eBusiness in turn. Along the way, we explain the essential technology, and give advice on its use (backed up with examples and case studies for illustration).

One point that will become clear as we proceed is that there is a lot more to eBusiness than shopping. It may be true that, to most people, the whole idea of eBusiness is new and trendy, an exciting opportunity brought about by the convergence of IT and the Internet. However, this is only half the eBusiness story. The other half concerns the supply chain and how traders are supplied with the goods and services that they need to operate. In the US, the term 'Maintenance, Repair and Operations' (MRO) has come to be used to describe the main commodity products used by a large company; and this side of eBusiness has been going on for some years.

The established mechanism for inter-company transactions is known as Electronic Data Interchange (EDI). There are established standards (called the EDIfact standards), and an established body of know-how in how to link to the fulfilment, inventory and other back-end systems that make it possible to automate the end-to-end eBusiness process, from ordering through to delivery. The 'behind the scenes' aspects are an important part of the overall picture, and will be explained fully as we unfold our story.

1.3 WHO SHOULD BE INTERESTED?

There are various categories of people who benefit from doing business online. The initial spotlight has fallen on the consumers, who can shop (globally, at all hours) without having to travel, can window shop to their heart's content and, in some cases, can have their needs directly fulfilled (especially if they are after software, games, music, *Which?* Reports, papers, legal advice and the like). There are others. In fact, there is a surprising number of distinct parties who can be identified. For example, the reasons why a supplier would be interested in eBusiness are quite diverse, and include:

- The small shop that wants to establish a global presence (such as Trophy Miniatures, a manufacturer of toy soldiers and probably the first multinational corporation with only one employee!). [http://www.trophyminiatures.co.uk].
- Affinity groups (such as Barclay Square, where a set of established, large trading organisations wish to extend their reach by associating themselves with others).
- Large retailers (such as Tesco online, a UK-based supermarket chain who see an online presence as a means of cutting their costs).
- New businesses (such as Amazon.com, who have extended the traditional image of a bookshop by allowing the buyer to participate in the business through reviews, promotion of books, etc., and IQ Port (now, sadly, defunct—proof of the volatile nature of eBusiness) which creates a new market in online knowledge).
- Suppliers, who want to speed up and secure their business (for instance, supermarket stock records linked directly to wholesalers' delivery schedules to allow 'just in time' deliveries).
- Large corporates, who may sell product but see the main opportunity as extending their services by making them available to consumers online (such as BT's 'Friends and Family' service, that allows customers to set up their own calling preferences).
- Trade bodies, co-operatives and other groups of companies in niche areas who, by pooling their eBusiness presence, can achieve objectives that would be unattainable individually. For example, small companies may have very little purchasing power individually, but through eBusiness, may combine their requirements and rival corporate purchasing departments.

Ultimately, there is the virtual retailer. This would be a service that consists of no more than a front-end with no physical presence (and no stock)—just links to suppliers, manufacturers and customers; and the wherewithal to fulfil an order and charge for the pleasure! It was through careful stock control that the 1980s saw the advent of just-in-time manufacturing. The new millennium could well see the next step: just-in-time fulfilment, where the supplier keeps the customer happy despite owning no physical resources at all.

Each of the above will have a different set of needs and priorities. Banks, for instance, will be more concerned than most with security. A supplier of car parts may be more worried about building a catalogue that is quick and easy to use, or ensuring that the online facilities are linked to established stock control systems. Matching the available technology to the purpose for which it is intended is a theme we shall return to throughout this book. For now the point is that eBusiness is not one thing—it is a spectrum of capabilities and opportunities that need to be balanced.

1.4 ARE WE READY?

We have already said that eBusiness is the result of a global public information network (the Internet) and sophisticated IT technology that provides the applications to run over that network. Here we look at these two base ingredients, and consider their impact on a world in which the pace of change is ever quickening.

First, the Internet. After it emerged from its military roots, there have so far been two quite distinct generations of the Internet. The first was its role in supporting the worldwide academic network, where it provided, in essence, a set of interconnected computers running specialised applications for expert users. The second generation can be characterised by the now familiar World-Wide Web (WWW), where the underlying technology is less apparent and the prime perception is of a ubiquitous fabric that gives best-efforts access to a large amount of interconnected information. The authors premise is that eBusiness provides the next generation Internet—one in which it provides the infrastructure for an interconnected business world.

It is not difficult to see why we put this premise forward. Internet revenues have been growing at more than 100% a year (compared with growth in the basic telecommunications business of 5% per annum). All of the major players in the communications industry see it as their main source of future profit, and are busily investing both money and intellectual effort in its development. With most of the high street outlets offering free Internet services, there is little money to be made in simple connection and mail services, so the new revenue streams will have to come through premium services. To achieve these at a commercially acceptable level, the providers need to tackle issues of security, quality of service and reliability—the very attributes that enable eBusiness to flourish.

At the same time, demand for the Internet is increasing at the staggering rate of 1000% per annum (see Figure 1.2 for the growth in Internet hosts). By the year 2000, half of all the bandwidth in the world will be Internet traffic; the other half will be everything else (such as voice telephone calls). Take that growth rate out another few years, and by 2003 the Internet will be more than 90% of the bandwidth, and by 2004 more than 99%. Of course, extrapolating current growth can be misleading—it is unlikely that we will all be online by 2005—but even if these projections are somewhat inaccurate, the same conclusion can be arrived at but a few years later. The bottom line is that such a ubiquitous infrastructure makes for a huge marketplace.

It would be wrong to focus solely on the infrastructure that can carry eBusiness transactions to a huge community of interest. The processors, protocols and products that connect to the Internet play an equally important role, and it is very clear that just about every major player in the IT

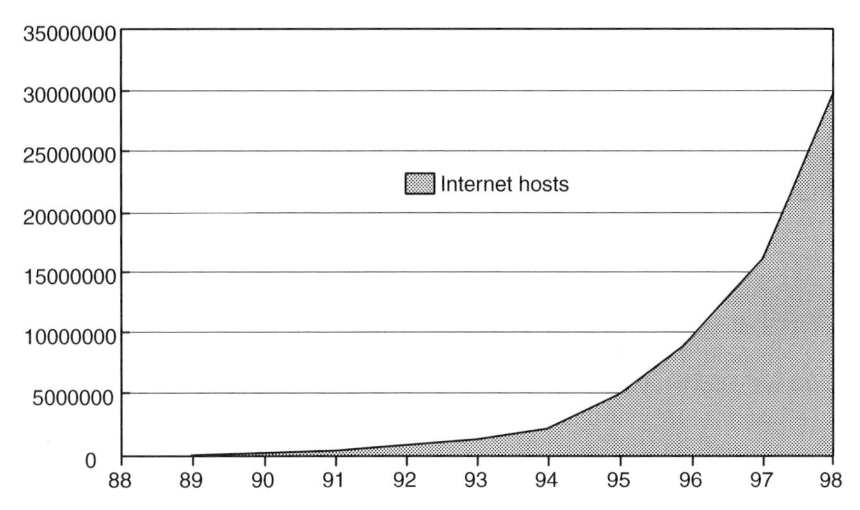

Figure 1.2 The inexorable growth of the Internet

marketplace has developed a product set that supports eBusiness. Some have been more explicit than others (for instance, IBM market specialist e-Commerce Services), but all provide some part of the jigsaw puzzle that is eBusiness.

It is clear, simply from looking at the strategic plans of a few of the major IT suppliers, that a rich infrastructure for eBusiness will develop. The lead in building a network-enabled IT world is being taken by the likes of Microsoft, Oracle and Cisco. By way of illustration, Microsoft's application architecture incorporates Internet connectivity as an implicit feature, Oracle's Web application servers are a central part of their product set, and Cisco are building security and quality of service as an integral part of their network devices.

So, whereas eBusiness technology used to be special purpose (i.e. with specialised equipment such as Point Of Sales (POS) terminals, specialised networks such as those run by credit card companies, and specialised standards such as EDI), it is now mainstream distributed systems and network technology. The basic needs of eBusiness are built into the thinking of all the key suppliers, and will increasingly be an inherent feature of their products. High-speed digital networks, powerful computers and standard software are heralding the death of distance.

Of course, having a fast, secure network and some slick applications doesn't mean that we all start using it. History is littered with good ideas that, for some reason or another, were never adopted. The signs for eBusiness are good, though. A basic reason behind this is that it is widely predicted that knowledge rather than location will define the corporate landscape of the future. There is substance behind this statement: a survey by Andersen Consulting and *The Economist* showed that 40% of executives believe that their organisations will be virtual by 2010, compared to just 3% who reckon that they already are.

Personal experience further backs this up—one of the authors gave up his office, desk, fixed telephone and (as far as he could) paper in the early 1990s. The experience was one of initial frustration and subsequent fulfilment. In the intervening years, the world has noticeably moved on to accommodate the virtual worker—location has indeed become less of an issue, and it is much easier to eschew fixed location in favour of a roaming brief. A major factor in this has been the way in which the technology has matured to allow fast, ubiquitous access to information. Exactly the same technology fulfils the needs of eBusiness.

Are we ready—you bet we are!

1.5 ABOUT THIS BOOK

Overall, this book will look in some detail at each of the elements of eBusiness, as illustrated in Figure 1.1. So far, we have done no more than say what each of these elements is. The next chapter starts to deliver something useful by examining the electronic marketplace in more detail. In doing this, we introduce the key players, their roles and how they interact with each other. There is a sting in the tail of Chapter 2, which is that despite most online trading models having a physical dual, some don't. How these new entities operate and evolve remains to be seen.

Having established the nature of the marketplace, Chapters 3 through 9 add much of the underlying detail and practical flavour for a real eBusiness implementation. In them we cover the mechanisms, ideas and techniques to cope with the issues of security, catalogue design, presentation and billing discussed earlier. Chapter 3 starts the journey in the practicalities of eBusiness by looking at catalogues—the basic mechanism for advertising your wares, online. The next chapter explains the payments mechanisms being used for Internet-based trade, and explains some of the accepted standards, such as SET, in some detail. Chapter 5 deals with the associated issue of security. We'd tell you more about it here, but it's a secret.

The focus changes a little in Chapter 6. Here we are concerned with 'integration', by which we mean the techniques for hooking existing (or legacy) systems and data into an eBusiness. This is an issue that has long vexed the designers of networked systems, but one in which there is now a wealth of experience and know-how. Despite being at the unfashionable end of eBusiness—hidden from the limelight of the shop-front—integration is crucial to successful implementation. In Chapter 7, we continue to address the less glamorous side of our topic by looking at supply chain automation. In other words, how one business trades with another on an everyday, high volume (and sometimes, mundane) basis.

Chapters 8 and 9 are very straightforward guides, respectively, to setting up an eBusiness and to the technology that is available. By the end of Chapter 9 it is intended that you should know what is available and how to put together an effective and practical eBusiness.

Given that the best way to learn about things is to see how they are done for real, the disparate points raised in Chapters 3 though 9 are brought together in Appendix 1, which contains some case studies. We have tried to pick examples that illustrate different aspects of eBusiness. More practical guidance is contained in the Appendix 2, which shows how to select the right technical approach for your preferred style of working.

To close the main part of the book, in Chapter 10 we take a peek over the horizon at where eBusiness is headed. Some of this is to do with what technology will make possible, and there are significant new capabilities just around the corner. To focus on the technology, though, would be to let the tail wag the dog. We also take stock in this chapter of the likely beneficiaries of eBusiness; in short, who will make the money.

1.6 SUMMARY

In a world where networks, software and computers provide the nervous system that any information-intensive business needs to compete, it should be no surprise that electronic business is seen by many as a natural progression in the way that we buy and sell. This chapter has introduced the notion of electronic business, and has described its constituent elements—a shop-front, buyer and seller catalogues combined with some mechanism for billing, settlement and fulfilment. In addition to outlining the technical requirements and options for eBusiness, we have also taken a look at the nature of the electronic market, and considered some of the 'look and feel' issues that make systems usable and appealing.

The picture we paint is one of mature technology enabling online business, and the growth of that business driving further advances in the capabilities of network and communication technologies. In particular, the growth of the Internet along with the greater sophistication of applications that use it, give us an impressive toolbox with which to build the online market.

In reality, of course, the networked application that fulfils all of the needs of the information age is a rare beast. Most of us still use the telephone as our primary means of communication, strain our eyes reading distorted fax orders, travel to meetings, and struggle with long and convoluted electronic mail addresses before resorting to sending discs via surface mail (an option affectionately known as 'Honda Net' or 'Frisbee Net'), because file transfers are too slow and tortuous.

There are two main messages in this chapter. The first is that eBusiness is unlikely to be a passing fad—it offers genuine advantages in terms of speed, flexibility and the removal of cost from the supply chain. From a business point of view, it is the key to competitive advantage. The second message is that an eBusiness is not a commodity purchase—to succeed requires a bit of thought, and a fair bit of know-how. The purpose of the book is to provide enough information so that you can apply business judgement to what is essentially a new business opportunity.

So, we have made a start in explaining some of the basic commercial and technical issues. The chapters that follow are aimed at building a balanced and detailed guide for those who wish to take advantage.

FURTHER READING

Atkins J & Norris M, *Total Area Networks*, John Wiley & Sons, 1999.

Cronin M, *Doing Business on the Internet*, Van Nostrand Reinhold, 1994.

Elsworth J, *The Internet Business Book*, John Wiley & Sons, 1994.

Frost A & Norris M, *Exploiting the Internet*, John Wiley & Sons, 1997.

Naisbitt J, *Global Paradox*, Nicholas Brealey Publishing, 1994.

Norris M, Muschamp P & Sim S, The BT Intranet—Information by Design, *IEEE Computing*, March 1999.

Negroponte N, *Being Digital*, Hodder & Stoughton, 1995.

Ohmae K, *The Borderless World*, Harper Collins, 1990.

Resnick R & Taylor D, *The Internet Business Guide: Riding the information superhighway to profit, SAMS*, 1994.

The Standish Group International, *The Internet Goes Business Critical*, http://www.standishgroup.com

The Consumers Association, On-line Shopping, *Which?* January 1999.

2

The Electronic Marketplace

Federations work better than monolithic organizations because, along with strength, they offer the degree of flexibility we need to deal with these turbulent times

Warren Bennis

In the business world, there is very little new under the sun. People have traded for thousands of years, and the basic model of two parties trying to satisfy each other's needs holds to this day. This simple fact is sometimes clouded, though, and the current cause of obfuscation is the use of networked computers to support the process.

We should say that the Internet has certainly made a big difference. It has opened the way for a new, completely electronic economy with different operational models, strategies, structures, politics, cultures, regulations, successes, failures, opportunities and problems. According to the Organisation for Economic Co-operation and Development (OECD), the economic performance of countries like Canada, the USA and the UK now compares more favourably with that of former leaders Japan and Germany, and a significant part of the reason for this is their aggressive exploitation of information technology.

It is not just national economies that have embraced electronic business. Many companies have seen tangible benefit. The IT manufacturer Cisco sell the best part of $10M per day via electronic commerce. They reckon that they save nearly $300M a year in staff, software distribution and paperwork costs.

Given this, it should come as little surprise that many pundits are predicting great things for eBusiness. Indeed, technology watchers Forrester estimate that electronic trade will grow to over $300Bn a year over the next three years. That is not to say that there are fortunes waiting to be made by anyone who gets their wares online—West's law of e-conomics states that for every business trying to make money out of the net, there is an equal and opposite business who will do it for free!

Of course, like any market, the only traders who will capitalise on the opportunity will be those with the right ideas, the right product set, in the right place with the right presentation at the right time. When you cut away all of the hype, there is much that remains unaltered and many established ways of doing things that work just as well in the electronic sphere as they did in medieval times. However, when things are done faster, over greater distance and with fewer human interventions, there is the prospect of a quick profit.

In this chapter we look at the electronic market and analyse its emerging structure. In particular, we look at the various ways in which ebusinesses can be conducted and, more to the point, how their structure bears on how they should be set up and the way in which technology is best deployed. In this respect, we will examine the goals of several different flavours of eBusiness, the typical flow of orders for each one, and key implementation issues such as fulfilment and payment.

2.1 WHAT IS AN ELECTRONIC MARKET?

An electronic market can be viewed as a direct parallel of the familiar shop, store or emporium. It is, in essence, a virtual trading area where deals are struck through a computer screen, over a network. The 'shop-front' is usually a set of web pages, the shelves equates to the catalogue where products are stored and displayed, and the warehouse is the server. In fact, there is an electronic analogue of all the items you'd find in a conventional market—the perils of fraud and shoddy goods go hand-in-hand with all the glitz and glamour.

To counter the less-than-welcome aspects of eBusiness, there are mechanisms to make sure that people are who they say they are, that cash changes hands as it should, and that goods are delivered. These mechanisms are becoming increasingly sophisticated, and can be complex and expensive to deploy. So it is important to determine exactly what is necessary for your particular brand of eBusiness. This means establishing a clear business model, one that fits the way in which you intend to operate and which meets your trading objectives.

The form of eBusiness familiar to most people is straightforward Internet shopping. More often than not, the goal of the trader is simply to exploit another (new and sexy) channel to market. In this instance, their aim would be to match the established shopping experience as closely as possible. Hence the interaction between supplier and customer would be something like this:

Customer	Message Flows	Supplier
Select suppliers web page		Establish shop front
	Request for goods →	Solicit details of the order Credit card validated Goods availability checked
	Order confirmation ←	
Enjoys purchase	*Order fulfilment* ←	

This is, of course, an oversimplified picture. The interactions with a bank to validate the customer's credit card and (possibly) with subcontractors to arrange for the delivery of goods are complex and not shown here (but are explained in some detail in later chapters).

The fundamental reason why the picture is a simple one is that the supplier doesn't really have to add very much—as stated above, this may be a simple extension of existing channels to market. In some instances, the supplier will do no more than mediate between the manufacturer and customer. An example of this would be Amazon.com, who provide a shop front for books but do not hold any stock, and are neither publisher nor bank. Hence, all activities other than presentation and marketing are done elsewhere.

A slightly different picture emerges if the supplier has some measure of control over the product being sold—either the ability to configure it or complete governance of production. This time, the sequence of interactions between customer and supplier could be:

Customer	Message Flows	Supplier
Select suppliers web page		Establish shop front with user configurable applications
	Product enquiry →	Presentation of options
	Select preferences →	Capture customer choice Check build schedule Validate credit card
	Order confirmation ←	
Enjoys choice	*Order fulfilment* ←	

The main difference in this set of interactions is that the supplier is presenting a product that can be customised, at least to some extent. Hence, the online experience is often extended to allow the customer to experiment with the basic product and see what it looks like in a different colour, configuration, size or style. This model works well for car manufacturers such as Daewoo, computer vendors such as Dell and fashion retailers such as Next and Top Shop.

The link between these suppliers is that they succeed by selling a basic range of products that can be tailored to suit a variety of customers. Allowing the customer to 'build' his or her own computer, car or suit using a web application provides a valuable extra capability at little cost. Furthermore, the door is opened for the manufacturer to gain customer information directly, without the need for a car showroom, retail survey or high street shop.

A third distinct model that is emerging in eBusiness is one where the prospective buyer submits a bid which may be accepted by the supplier—rather like a traditional sealed bid auction. In this case, the aim of the eBusiness is to match a customer who wants something with a supplier willing to sell it for the price that has been bid.

As with the first model, nothing is actually changed about the product being sold—the idea is to provide the manufacturer with a means of shifting their product for an optimum price. The sequence of interaction is rather like this

Customer	Message Flows	Supplier
View catalogue		Maintains catalogue
	Submit bid →	Collect all bids Select winning bids Match buyers and sellers
Enjoys bargain	*Bid accept/reject* ←	Collect commission

This type of model works well with time critical and low transaction fee products such as hotel and aeroplane seat reservations. The eBusiness stays solvent by taking commission on every completed transaction.

To recap, we have introduced three basic operating models that can be used with eBusiness:

- The broker model, in which the eBusiness may or may not have tangible presence. Its aim is to market a predetermined set of goods and services. A good business is distinguished by the following key features—attractive packaging, efficient delivery and accurate payment handling.

- The customisation model, where the objective is more likely to be the selling of a customisable range of products to a mass market. In this instance, the business should have online applications that enable prospective buyers to adjust the basic product to their specific needs and preferences.
- The contact model, where the aim is to be a go-between, matching prospective suppliers of goods and services with buyers. Typically, this is a high volume business with speed and efficiency of all electronic transactions being the key point of differentiation.

There are a number of technical issues that all of the above models should be concerned with—security and payment being the two most critical. These are complex and detailed subjects that need be addressed in some depth, so will be covered in detail in later chapters. The key point for now is that each model has its own success criteria and technical requirements, so it is vital to be very clear in which model you intend to use.

It would be nice to think that our three models covered all eventualities in the electronic market. Regrettably, life isn't that simple. There are some more subtleties that we need be aware of and consider, so let's take another viewpoint on trading in the electronic market.

2.2 TRADING MODELS: MARKETS, GUILDS AND AUCTIONS

So far we have established that the nature of eBusiness lends itself to more than one trading model. The three basic types introduced above were the broker model, the customisation model and the contact model. Each can take a number of different forms. By way of example, the broker model can be realised, quite simply, as an additional (electronic) channel to market for a retailer. Anyone can set up some web pages and start selling over the net.

The broker model fits equally well when applied to the age-old idea of a marketplace owned by one organisation, populated by a number of (authorised) traders. In this case, the owner of the online presence doesn't necessarily provide any of the goods and services for sale. Barclay Square is an example of this—the offerings of a number of suppliers are made available though Barclays' eBusiness set-up.

If we extend this same model with the idea of having guilds that control standards, we have the style of trading pioneered by companies such as IQ Port. In this instance, the traders in the marketplace are organised into (and under the auspices of) a guild that ensures quality. This is a direct copy of the medieval marketplace, where a guild would train an individual to be a Silversmith, Fletcher or Cooper who could be trusted to uphold the values of that guild.

The point is that there are many variants on the basic theme. In fact, there are a whole host of ways in which a marketplace can be characterised. To do this for the electronic market, we need to consider what the key influence on the market are and, from observation of evolution to-date, the three most significant appear to be:

- *Type of participant*. The familiar view of eBusiness is of business-to-consumer transactions, but business-to-business is at least (if not more) significant, and there is also the prospect of consumer-to-consumer dealing.
- *Importance of timeliness*. Does the transaction have to conducted in real-time, somewhere near real-time, or can it be more like an electronic mail interaction?
- *The driver for the marketplace*. Is it dominated by the suppliers or by the buyers, or does neither hold the upper hand?

It is interesting, and quite useful, to look at the options described in the last section against the some of these characteristics. Taking the last bullet in the above list, along with the earlier trading models, allows us to construct a two-dimensional view of the market. The first dimension is the dominant drive in the market—whose brand attracts participants? The second dimension is the nature of what is being advertised—is it mostly offers from sellers or requirements/bids from buyers?

In Figure 2.1 we have taken the three main options on each axis, and identified some examples that fit into the resulting matrix. The options used on each axis of the figure are worthy of a little explanation.

On the 'drive' axis:

- A seller-driven market is deemed to be one where a small number of suppliers with strong brands can attract a larger number of buyers into their own marketplaces.
- A buyer-driven market is one where one or more strong buyers can attract their (usually many) suppliers to participate in a marketplace for the exclusive benefit of those buyers.

On the 'lead' axis

- An offer-led market is deemed to be one where it is the seller that advertises their products or services. The broker and customisation models introduced earlier on would fit in here.
- A requirement-led market is one where it is the buyer that advertises his requirements.

Drive - (whose brand attracts participants?)

Lead - (what is advertised?)	Seller	Marketplace	Buyer	
Offer	Supplier with strong brand in their market, publishing a catalogue. Direct banking	Suppliers without very strong brand, publishing product catalogues in a strong marketplace, Auctions	Hub & spoke model focussed on very strong buyer(s), incorporating product catalogues	Offer-led markets
Either		Classified adverts, Spot markets	Hub & spoke model incorporating tenders as well as product database	Jointly-led markets
Requirement	Supplier with strong brand selling customised products	Service allowing buyers to issue RFQs for products and services	Very strong buyer issuing ITTs	Requirement-led markets
	Seller-driven markets	Open markets	Buyer-driven markets	

Figure 2.1 Dimensions and examples of the electronic market

In addition to these fairly polarised options, there are open markets in which there is neither a strong buyer nor a strong seller that attracts participants.

Figure 2.1 can be specialised depending on a number of other relevant characteristics of the marketplace—such as whether the supplier is primarily focused on business or consumer customers. In addition, it can be useful to see where specific types of market fit onto the figure. For instance, it is quite easy to see that consumer-to-consumer marketplaces will be open markets, since neither the buyer nor seller will have a strong brand in their own right.

Just as in the real world, the majority of today's trading applications are seller-led. Any supplier wishing to reach out into the online marketplace can readily set up an independent online store to market their products or services. However, the new online medium is also introducing new ways to find information, suppliers and product advice. This is causing a fundamental shift in the relationship between buyers and sellers, with the potential of empowering the buyer rather more than has been possible previously.

The sheer flexibility of the electronic market will, most likely, lead to new ways in which business can be conducted. For instance, there will probably be significant growth in open markets, operated by intermediaries. This model seems particularly relevant for commodities with complex requirements and specifications, such as insurance and holidays. It might not be very long before a crofter in the Highlands of Scotland sorts out your bungee-jumping holiday in New Zealand!

In addition, there is an emerging opportunity to provide procurement solutions for major corporations who are motivated by the need to control their total purchasing expenditure and rationalise their supplier base. This new application will primarily address indirect goods, otherwise known as Materials, Repair and Operations (MRO) spend, and will be tightly integrated into the mainstream Enterprise Information Systems, such as SAP, Oracle Workflow, Peoplesoft and BAAN, being used to handle basic business functions in large organisations.

One general point to make is that the electronic marketplace both admits new players and cuts out existing ones. In the former case, the growth of open markets will spark new roles. In the latter, direct sales seem likely to cut out many retailers and middle men—an effect termed 'disintermediation'.

Whatever the possibilities and speculation, the way in which the electronic market develops will be driven by the people that use it. So, in the next section we revisit and examine in some detail our buyers and sellers by considering the trends and developments in each of the three vertical columns shown in Figure 2.1.

2.3 SELLER-DRIVEN, BUYER-DRIVEN AND OPEN MARKETS

Markets are often categorised by the dominant party, and there are three quite distinct types in eBusiness—seller, buyer and open:

- In seller-driven markets, it is the large, dominant retailers and manufacturers who set the price. Their brand or product is sought after, so its price is not really for negotiation.
- In a buyer-driven market there are many people selling into the market with similar or competing goods, and only a small number of dominant buyers who are looking to get the best deal.

These dynamics have considerable impact on the appropriate technology for eBusiness, specifically, the type of catalogue that is used. In the seller-driven market, a hub and spoke catalogue—illustrated in Figure 2.2—fits well.

BT TradingPlaces is a good example of a centralised catalogue system. In this case, suppliers enter product and service information onto a central system that the buyers then search. The focus is very much on the central

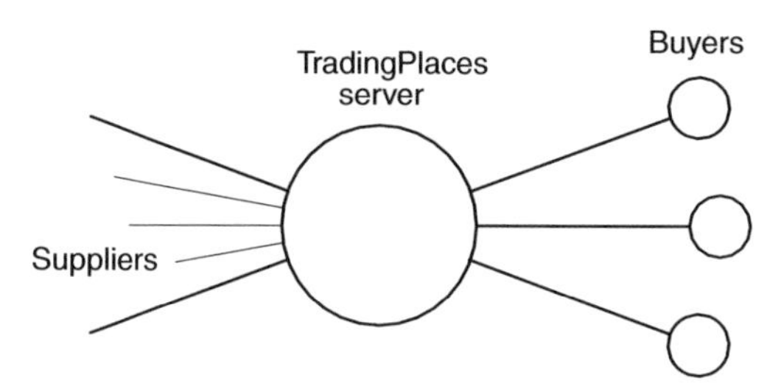

Figure 2.2 A centralised catalogue

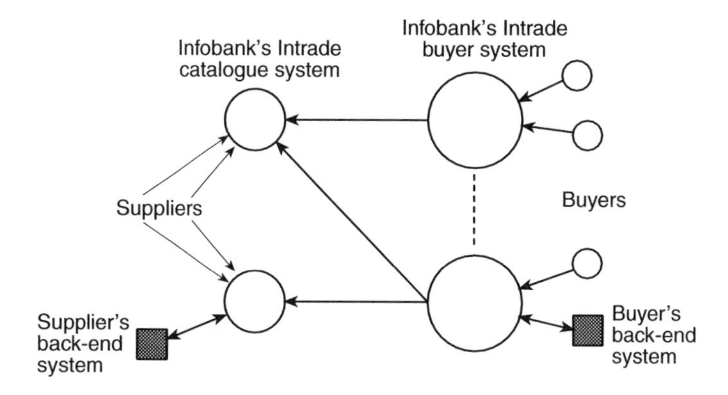

Figure 2.3 A distributed catalogue

data store, which is controlled, managed and secured.

In the buyer-driven market, a retail catalogue is more appropriate and a different architecture, more oriented to giving access to many suppliers, applies. This is shown in Figure 2.3. Instead of having one, big (supply) catalogue, the distibuted architecture (of which Infobank is an example) takes feeds from a variety of suppliers and aggregates the information gleaned into a form that the buyers can use.

There are also open markets in which the buyer and seller negotiate, or a free market, where the behaviour of the market itself sets prices. Insurance brokering is an example of this, as are optimisation packages for selling airline seat capacity.

We now examine each one in detail.

Seller markets

Today, most companies setting up to trade online and service their customers through this new channel to market do so unilaterally. In almost every case, the supplier will be advertising products or services in some kind of catalogue. Consequently, these applications will mostly belong on the seller-driven, offer-led segment of the marketplace. One of the major differentiators in this space will be whether the application is to be focused on a business customer, a consumer or, as can often be the case, both business and consumer.

Business customers will only use the online medium if it offers significant advantages over traditional means. They will be motivated to be able to transact quickly and efficiently, and may well be put off by complex sites offering a 'shopping experience'.

It seems inevitable (at least, if they are to work) that business-to-business trading models will become more standardised—automating the interface between the buying and selling organisations. This will inevitably mean that trading environments aimed at business-to-business markets will need to adopt international or industry-wide standards as and when they reach a critical mass of adoption. Suppliers will generally be focussing on providing service to their existing accounts, and will need to be able to reflect customer-specific pricing and service to individual buyers.

As buyer-driven Maintenance, Repair and Operations (MRO) procurement systems are deployed, suppliers may need to be able to establish a presence on a number of their customers' MRO systems, as well as maintaining their open Internet presence. How this can be achieved will depend upon the architecture of the procurement system. More on this later.

In contrast to the business-to-business side, business-to-consumer applications will be competing for customers in a number of new and imaginative ways. It is far more likely that consumer-focused businesses will commission bespoke developments in order to differentiate themselves and incorporate new capabilities ahead of their competitors. A bit of gloss, sex and seduction helps in this instance.

One of the crucial factors affecting the success of a trading application, especially for small and medium enterprises without a really strong brand, will be how effectively new buyers are attracted to the site. Today a supplier will submit details of their site to the different search engines, and may commission advertising banners and directory listings at Web portals and other strategic web sites. However, the search engine approach is somewhat 'hit and miss' from a buyer's perspective. There is a relatively low likelihood of a search engine returning a meaningful result to a query because the HTML that makes up a web page lacks any mechanism to encode meaning about the web page. Moreover, search engines are often unable to deal effectively with dynamic data-driven web sites which increasingly form the basis of online catalogues.

An emerging solution to this problem is the advent of XML (extensible markup language), discussed in some detail in Chapter 7. Once widely adopted, this will allow search-engines to 'understand' the nature of the content of a catalogue rather more effectively than at present, and therefore allow buyers to locate products (and suppliers) rather more easily than at present. Once search engines are able to provide more meaningful product searches, they will start to become the 'front-door' through which a buyer can begin their search for suppliers—they will in effect be well positioned as open marketplaces (see below). BT spree is an example of this.

Buyer markets

Another class of trading environment is one in which an organisation has sufficient power as a buyer that they are able to dictate to their suppliers that in order to trade with the buyer, they must participate in the buyer's own 'marketplace'. This participation will extend to the publication of their product portfolio into a particular catalogue format, and the acceptance of purchase orders via the marketplace. This is a similar phenomenon, to that experienced with the roll-out of EDI via the so-called 'hub and spoke model', shown in Figure 2.2—a major purchaser could dictate the terms under which they would do business with their many but smaller suppliers.

Goods purchased may be broadly grouped into two types:

- business-critical procurement such as raw materials and components for a manufacturer or products for a retailer;
- non-critical items used for Maintenance Repair and Operations (MRO) such as stationary, IT equipment, etc., also known as 'goods'.

The requirements on systems to support these two types of purchase are slightly different. In the first case, procurement is specified by a small group of specialised buyers, perhaps assisted by Materials Requirements Planning (MRP) systems which need to be able to source goods reliably, and quickly identify alternative suppliers in the event of a problem with the preferred supplier. In the retail sector, access to a broad supply base is particularly important. For MRO goods, on the other hand, purchases will be originated from personnel throughout the enterprise, and one of the major challenges is to rationalise the supplier base for particular products and drive down 'rogue' purchasing to maximise the ability of the enterprise to negotiate favourable trading terms with the supplier.

The drive here is for the buying organisation to be able to significantly reduce the cost of their MRO expenditure, as well as to be able to accurately monitor and, if necessary, control expenditure of this kind. It is anticipated that the opportunity for providing MRO solutions will be greater than that for business-critical procurement.

Although the simplified view here is of a trading environment focused on serving one supplier, there will be a very natural migration to open the environment up to other purchasing organisations who deal with the same supplier base. In this way, a marketplace that starts out as being conceptually driven by a single buyer will very quickly become an open marketplace, in the sense that it supports multiple suppliers and buyers.

Open markets

From the earlier discussion, it is apparent that suppliers adopting seller-driven catalogue applications and purchasers investing in buyer-driven (e.g. MRO) applications will naturally tend to become participants in open marketplaces in some form or other, even if this is not a conscious decision. Moreover, it will become increasingly important for a supplier to be able to publish product data into a stand-alone catalogue, as well as relevant open-markets.

Open markets may be private (supporting a closed buyer community) or public. The private markets will typically be characterised by services targeting business MRO procurement. In this instance, a significant level of investment in an Intranet procurement application and the associated integration into back-office systems will be necessary in order to participate as a buyer. Public markets will most likely emerge from community of interest Web sites, extranets and search engine portals.

Private open markets

There are a number of system developers such as Ariba, CommerceOne and Infobank developing an MRO procurement proposition to take to major corporations throughout the world. The proposition is that once a company has an Intranet in place, the Intranet can be used to deploy an application to be used by anyone within the organisation who wishes to procure MRO type goods. The application can implement the processes used within the company to authorise requisitions, place the purchase orders and control the allocation of budgets, as shown in Figure 2.4.

Bringing this information together means that rogue purchases can be brought under control, and the company can streamline its supplier base—increasing its ability to negotiate favourable trading terms. More automation of the procurement process is also an objective: for example, the UK Ministry of Defence has identified that it has spent over US$ 120 to purchase a US$ 1.60 padlock, as a result of the paperwork and time spent on raising the purchase order. This is by no means an isolated example, and there are many organisations who have handling charges that exceed the cost of the item being handled. Systems such as Commerce One's Market Site and Buy Site claim savings of as much as 90% from the overall cost of processing orders.

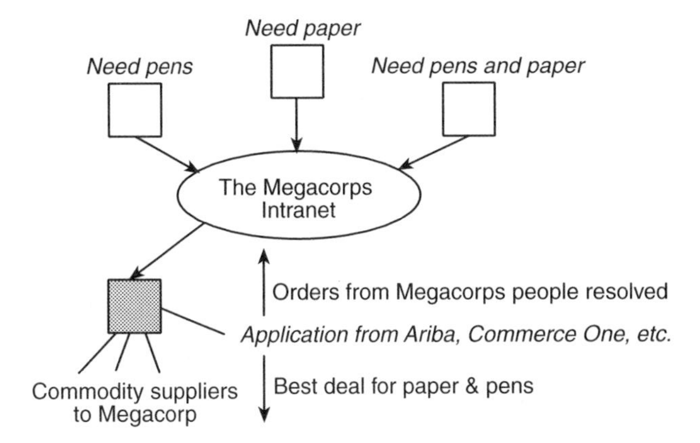

Figure 2.4 A third party looking after commodity supplies

The other major part of the proposition is that the system can be used to deliver an online product catalogue for each of the approved suppliers to the desktop of everyone in the organisation, so that employees have up-to-date product information available to them when required. The mechanism for keeping a purchaser's Intranet catalogue in step with a suppliers main data source all vary slightly from one implementation to another.

Public open markets

Although buyers are readily able to find suppliers with well-known brands, it will still be a very laborious process for a buyer to investigate a wide range of potential suppliers without some kind of intermediary service working on the buyer's behalf. In fact, although it has been a commonly held view that the Internet will lead to substantial disintermediation, there will be significant opportunities for new intermediary services perceived to be acting on a buyer's behalf (Carnall 1995). These new services will have the opportunity to become 'portals' through which buyers can access online suppliers. This trend is already starting with the major search engines offering shopping services; for example, see Excite shopping (http://www.excite.com/shopping/). Other players will specialise in specific vertical sectors.

Advances in technology, particularly the use of agents, will arguably have the greatest impact in this domain. A consumer deciding that he wants to make a purchase will very often not be an expert in that domain, and will therefore be receptive to assistance in deciding precisely what product to buy and where to buy it from.

Imagine someone wanting to buy a computer, but finding the whole subject quite bewildering. He may have a very good idea of what he wants it for, e.g. to run an accounting package, word-processing and producing advertising brochures—but no idea of how this translates into what hardware and software to buy. To help this buyer towards his goal, an intermediary service could provide an advisor agent to help translate his need into a product specification. This agent role has been termed a 'generic advisor', and a good illustration of this approach is PersonaLogic (http://www.personal-logic.com).

Secondly, the buyer will need to decide which manufacturer and model he should buy. It is envisaged that a second type of agent, a 'specific advisor' (Carnall 1995), will be able to take the specification and provide the buyer with information from independent product reviews, electronic word of mouth and, possibly, the views of other similar purchasers to help him make this decision. Excite Shopping is now moving towards providing this kind of service for certain product areas.

Finally, the buyer needs to find the best supplier for the particular products that he has identified as being those he wants to buy. There are a number of approaches being adopted by prospective open market services here to provide comparison shopping such as Anderson Consulting's early BargainFinder agent, or Jango as offered by Excite.

An alternative approach is for the buyer to lodge a Request-For-Quotation (RFQ) with the service—the same approach as used in tendering for major contracts, but using the Net to scale the concept for the mass-market. In Figure 2.1 this approach corresponds to the bottom row labelled 'requirements-led' market. The efficiencies for the buyer come from having to define his requirement only once, and not having to know about all of the prospective individual suppliers. The service will send this requirement (now a sales-lead) to suppliers, who then have the opportunity to respond with a competitive offer. A good example of this approach is the Auto-By-Tel service, which specialises in providing a prospective car buyer with quotation from several dealers (http://www.autobytel.com). This approach can result in lower costs to buyers, as a result of the suppliers having a lower cost of finding customers—they do not need to advertise, as the sales-leads come to them from the intermediary service.

In the future we envisage that the distinction between RFQs and product finder agents will disappear from the user's perspective—he will provide the intermediary service with a specification, and then corresponding offers will be returned perhaps using both mechanisms to interface to the supplier. Also, it is apparent that all of the discussion in this section has focused on the activity of sourcing the product rather than the mechanics of conducting the transaction. This is deliberate, because it is felt that the primary focus of open-markets will be in empowering non-expert purchasers to make very smart purchasing decisions.

2.4 COTTAGE INDUSTRIES AND GLOBAL MARKETS

Establishing a clear and consistent picture of the market and your role in it is a necessary but not sufficient pre-requisite of a successful eBusiness. In addition to knowing how you should interact with others, you need an overall strategy and a detailed picture of what technology is appropriate.

One of the first questions that needs to be answered before jumping on the bandwagon is whether the eBusiness is intended to be an integral part of future activities, or whether it is intended to be a complementary add-on. If it is the former, it is a good idea to examine established processes (such as inventory control, accounts payable, purchase orders, etc.) to make sure that electronic transactions will work with the manual ones (and that some benefit actually accrues). The latter is unlikely to pose many problems with established processes—maybe some dual keying and/or copying of electronic orders, etc.—but it still needs to be examined for viability.

To illustrate this last point, in the US there are about three million people employed to process the paperwork in inter-business transactions (sales supervisors, shipping clerks, etc.). The cost of these 'transaction workers' could well be as much as $250Bn in the US (and probably over $1Tn worldwide), but the purchase order part of transactions is only about 7% of the total cost. So for eBusiness to really cost in, it would have to deal with selling, shipping and receiving, as well as purchasing. Hence, an examination of the end-to-end processes can indicate where the use of new technology with the existing business processes (most often, those that are not amenable to change) might fail to deliver significant benefit. Having the scope to install processes that fit with electronic trading is often a key factor in deciding whether eBusiness is appropriate.

There are a host of other questions that should be considered. For instance, what is the end game; to increase current business by extending channels to market, to make the business more efficient by cutting out manual processes, to diversify into new areas that are well suited to existing core skills and online trading? A little thought on just what strategy is being followed is useful, in that it helps in slimming down a wide range of possible options.

One thing that it is important to realise is that some of the traditional factors that need to be considered in relation to market strategy simply don't apply with eBusiness. For instance, size really doesn't matter. It is quite possible to conduct a large volume of business over a wide area with little overhead, irrespective of the trading model that has been adopted. Thus, the information smallholder can compete on an equal footing with the large corporation—in fact, it is not that easy in practice to distinguish the two.

It is not difficult to establish a framework of suitable attributes for each type of business. It is more of a challenge to come up with a definitive and enduring one—the whole area moves too fast and has too few established standards for this.

Nonetheless, it is useful to look at the degrees of freedom that do exist (and this is before operational dynamics come into play).

Distribution options can be

- Physical—direct fulfilment or logistics partner (such as UPS, FedEx).
- Electronic—downloading of software file, or music on to a device such as a Rio.
- None—if the product is access to information.

Application options can be

- One-to-one—a basic consumer/supplier interaction.
- One-to-many—for instance, a supermarket business with many produce suppliers.
- Many-to-many—mediation, as in the auction model, where contact between prospective buyers and sellers is key.

Payment options include

- Direct invoicing—the merchant gets their money by traditional means
- Third-party mediated—using electronic card authorisation and settlement (as provided by SET, etc.).
- Pre-paid Token—where a customer has an account that is credited by direct payment or earnings and debited by withdrawal or purchases.

Any particular eBusiness will deploy one of the options in each of the above categories. Given that there are 27 possible combinations and each of these is valid, there is a wide variety of technical support requirements. The authors wanted at this point to present a matrix of appropriate technical choices for each set of options. However, the nature of electronic business defies such a systematic approach, and each case has to be taken on its own merits.

Why this is so can readily be illustrated. Two similar eBusiness propositions may start from a completely different place. One company may already have its customer records in a database, the other may have all the records on file cards. The transition to an eBusiness will be viewed differently, with the former more inclined to seek more integration between the online shop and their back-end systems.

One last point on eBusiness strategy is that we have (so far) focused on new opportunity. There are sometimes equally persuasive reasons to go electronic based on defence. For instance, the UK government has set the target that by 2002, 25% of all dealings by both citizens and businesses with government should be conducted electronically. Already, discounts and incentives are available to those who comply. Penalties for those who cling to phone, fax and paper are probably on the way.

2.5 WHAT ARE THE ROLES OF THE PLAYERS?

An eBusiness that looks like a cogent entity to the consumer may in fact consist of many separate (but co-operating) parts. The network may be supplied by one company, content from another, application hosting (the clever facilities available on the user's PC and server) from another, delivery vans from the Post Office, and all branded by someone completely different. It is a more complex model than the traditional one of manufacturer who supplies wholesaler who, in turn, supplies retailer.

This section explains what an eBusiness needs to be an effective player—as well as the basic technology to enter the market, they need brand, content, service support, a means of fulfilment, etc.

When everything is connected, federations work better. Put another way, anyone who thinks they can conduct electronic business in isolation is probably in for a disappointment. The Seven Samurai waiting in the wings to help the prospective eBusiness (for a modest fee) are:

- *The commerce service providers.* These are the next generation Internet Service Providers (ISPs). Many of the large network providers, such as BT and AT&T, have been pushing their business up the value chain by offering a range of eBusiness applications and services. At the same time, the established ISPs, such as AOL and Demon, have extended their web hosting service to include integration of eBusiness applications.
- *Application providers.* Several of the major computer manufacturers, notably IBM, Compaq, Netscape and Microsoft, are offering application platforms that enable a merchant to build their online presence. In addition, specialists such as VeriSign (who supply digital certificates) have cornered the market on particular aspects.
- *Advertising.* Any busy marketplace attracts advertising, and the likes of Time Warner specialise in getting messages to the right place.
- *Logistics.* With the unavoidable need to move physical goods from suppler to customer, the leading logistics companies such as UPS and FedEx have linked their transport control into other people's order systems to give automatic fulfilment, along with real-time tracking of shippings.

- *Brand/security*. Some of the better known companies have set up online areas where traders can do business under the banner of a trusted name or brand. Barclays Square is an example of this.
- *Settlement*. In nearly all business, money changes hands, so someone needs to ensure settlement is effected between traders. The established banks provide online card authorisation.
- *Trade partners*. With the need to seek ways of offsetting costs and/or extend business reach, there will always be a need for trade partners and affinity business.

In addition to these players, there are others who play some part. Also, many of those mentioned above are seeking to extend their offerings to cover more of the needs of an online merchant—several will provide an eBusiness in a box (web pages, security, payments, the lot). The one thing that is clear is that as well as there being money in the transaction of online business, there is also the prospect of money in the setting up of eBusiness. The wealth of suppliers now popping up with eBusiness products and services is testament to this.

A final thought on suppliers is that many of them are themselves in the vanguard of online trade. One of the reasons that they are a good supplier is that they are in the process of taking the medicine they prescribe!

2.6 CULTURE, TRUST AND LOGISTICS

Whatever the style of trade that is followed, there are some inevitable issues that arise when the electronic world meets the real one. In particular, there are cultural, trust and geographic dispersal considerations that have to be attended to. The best eBusiness can falter if the fulfilment of an order costs more than the product being sold, if the shop window is unappealing or perceived as being from the 'wrong part of town' or the order form offends local sensitivities.

Culture

In principle, any globally accessible medium should expect to receive enquiries, orders, praise, etc. in equal measure from across the world. In reality this is far from true, and the reason can often be traced to the subtle but pervasive issue of culture. There are different ways of doing things in different countries, and accepted practice in one part of the world might be seen as strange behaviour in another. It is not feasible to present a face to the world that has uniform attraction, but some basic research can pay dividends

A good example of cultural difference is the national preference for particular goods or services. For example, many more Americans than Europeans are involved in share dealing, so a financial services company can reasonably expect to attract more business from the US than elsewhere. In this instance, a bit of cultural background may not have a great impact on the way the eBusiness is set up—but in other cases it just might.

Some of the cultural issues that are likely to have significant impact on the way an eBusiness is established and operates are to do with attitudes to payment and the perceived trustworthiness of an online trader. For example, in some of the Southern European countries there is a culture of face-to-face trading that leaves people less willing to make binding commitments over a network. Also, in some of the Eastern European countries, the use of credit cards is less widespread, as people prefer to deal in hard cash.

Given the diversity of world cultures and the speed of social development, it isn't feasible to accommodate all relevant issues, but they often do have some impact and should, at least, be included as part of the marketing considerations that preclude a headlong rush online.

One of the more tangible results of a culture is the law that it produces. In the main, there are few national laws that attempt to constrain or regulate eBusiness—after all, it is international by its very nature. Hence, for legal issues, it is worth looking at the likes of the United Nations. Their commission on International Trade Law (UNCITRAL) is the body most concerned with eBusiness. It has been working for some time on the consequences of electronic commerce and, in 1996, completed the Model Law which states provision for legislation (e.g. standards for Certification Authorities, rules for recognition of electronic signatures, mutual recognition of digital certificates).

There is, however, some way to go in this area before clear and useful statements can be made. For now, the assurances that certificates offer are underwritten more by the organisations that provide them (e.g. VeriSign) than by international law.

Trust

Very closely allied to culture is the issue of trust. When a web site can appear and then disappear (literally) in a few minutes, some measure of credibility and permanence needs to be established. Part of this can be achieved with a recognised brand—a familiar logo or a well-known product. Trustworthy payment arrangements (using digital certificates and trusted third parties) also give the buyer some confidence.

One of the most promising means of ensuring that online transactions are trustworthy is to implement a mechanism that proves people are indeed who they say they are. Digital signatures provide the same (if not better) level of assurance as a physical signature. These can be provided using public key, or asymmetric, cryptography. This is explained in more detail in Chapter 5, but here is a brief preview.

Public key cryptography systems use two keys—a public key and a private key. Each key (both of which are large numbers with special mathematical properties) has a specific purpose. The private key is kept a secret by its owner, but the public key is made accessible to all and sundry (it is placed on the certificate that is used as proof of identity, and is made available on a public directory). The keys can be applied in any order—a message encrypted with someone's public key can only be read by applying their private key, and a message encrypted with someone's private key can be decoded with their public key. Hence, public key cryptography can be used to provide a digital signature, as well as for encrypting files (Figure 2.5).

As we said, anyone with access to someone's public key can decrypt a message that has been encrypted with that person's private key. If the only person with access to that private key is its owner, then they must have 'signed' it for it to be readable.

There are a couple of other issues with public key cryptography that are very relevant to eBusiness. The first is that this mechanism of signing is rather stronger than the traditional signature. Because a signed message can only have been produced with the private key, the sender cannot deny that they sent it. So there is an extra layer of assurance with online dealing when the recipient is sure both that the sender is who they say they are and that they cannot deny that they sent the message. The ability to identify the originator of a message and make sure that they cannot deny having sent it is termed 'non-repudiation'.

The other issue with public key cryptography concerns the security of the keys themselves. These small secrets (that protect big secrets) are vital,

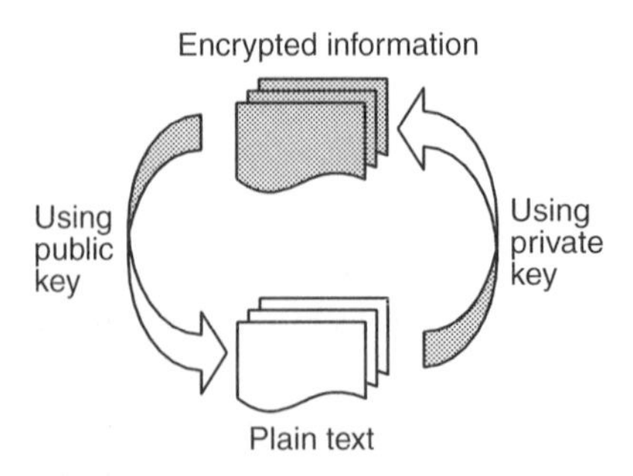

Figure 2.5 Using public and private keys to create a digital signature

and need to be looked after, so there is considerable sensitivity over their storage. At the same time, there is concern that organised crime makes considerable use of encrypted information (the UK serious fraud office encounters encrypted files in about half of its cases). So the issue arises of whether certain authorities should have access to keys. This is an, as yet, unresolved question, but one that will certainly impact on the rate at which eBusiness grows.

Guidelines from the Organisation for Economic Co-operation and Development (OECD) take the view that governments have a responsibility to promote commerce and protect privacy, while at the same time maintaining public safety and national security. Exactly where the balance lies has yet to become clear. In the UK, there was considerable debate over key escrow (whether the government should hold copies of keys as a security measure), the issue being one of too much protection placing a barrier on growth.

Logistics

By its very nature, eBusiness transcends established international borders, but the traditional mechanisms used by governments to protect their tax base and regulated industries are designed to deal with wholesale imports—containers of cars, wheat or paper. They are not geared up to deal with the likes of music, software and information, all of which can be delivered electronically. Just how regulation and tax will affect these low-cost, intangible items, delivered on an individual basis remains to be seen.

Online business is not confined to non-physical goods and services, so the question of physical movement of goods still needs to be considered. Logistics is already a sophisticated discipline, and many of the main distribution organisations have moved into electronic commerce. UPS, DHL and FedEx have all formed relationships with suppliers—the FedEx case studies (Appendix 1) go into some detail on just how readily the logistics organisations have linked into other people's eBusiness, and have come to provide natural partners for delivery.

2.7 SUMMARY

One of the notable features of eBusiness is that size does not matter—a sole trader can be indistinguishable from a multinational corporation. That does not mean, though, that everyone is equal in the eyes of the web.

There are various ways in which you can set up an eBusiness, and the way in which it is done will have major impact on its effectiveness. The key thing is to make sure that the style, technology, security, etc. are appropriate, and to do this, it is worth considering exactly what sort of market you are in and what sort of trading model you should be adopting.

In this chapter, three basic models are introduced—broker, customisation and auction. Each defines a way of trading on the net, and each has specific implications for the way in which the business should be set up. But this is far from the whole story. There are a whole host of subtleties that define the nature of the marketplace. In particular, the influence of a dominant player (supplier or customer) and the impact of trust, culture and logistics is considered.

Successful traders understand their market. This chapter has explained the nature of the evolving electronic market. As well as looking at the operational dynamics, the drivers and issues that shape the way in which business is transacted are explained.

REFERENCE

Carnall C, *Managing Change in Organisations*, Prentice Hall, 1995.

FURTHER READING

Fuglseth A & Gronhaug K, IT-enabled redesign of complex and dynamic business processes. *Omega*, **25**(1), 93–106.

Gore A, *Infrastructure for the Global Village. Scientific American*, 1995 special issue, p. 156.

Frost A & Norris M, *Exploiting the Internet*, John Wiley & Sons, 1997.

IDC research of European Internet commerce revenue, http://www.zdnet.co.uk/pcmag/supp/1998/e business/

Ohmae K, *Borderless World*, Harper Collins, New York, 1992.

Pine J, *Mass Customisation*, Harvard Press, 1992.

Shaw J, *Market Planning*, Artech House, 1999.

3

The Electronic Shop

Art is making something out of nothing and selling it

Frank Zappa

The first encounter that most people have with eBusiness is the electronic shop, by which we mean the 'look and feel' of the screen in front of them. Like any other business, there are a number of features and qualities that a successful shop must have.

First, it must look good—if it doesn't engage the attention of the viewer, then custom will just pass on by. This is a start, but by no means the end. The next thing that has to happen is that the viewer needs to be seduced in some way. There needs to be something in the electronic shop-front that holds the attention or, better still, triggers a purchase. Finally, there has to be some very straightforward way of completing a transaction. After all, the best way to sell is to make it easy to buy.

In many ways, the electronic shop is no different from its high street cousin, so it should be quite feasible to translate all the tricks of the trade and best practice from the world of retailing onto a computer screen. But to do this, we need to think about the goods that are being sold and how they should be presented. That is what this chapter is all about. It introduces the concept of the catalogue, and explains the various types, how they are implemented and how they work.

An important point that emerges as the chapter unfolds is that different types of eBusiness call for different approaches. For instance, a buyer with a large supermarket chain with many suppliers is not interested in pretty pictures of fresh vegetables. They want to be able to compare costs and delivery schedules across their suppliers. On the other hand, online fashion retailers need to convince the buyer that they are getting something truly desirable.

There are several different ways of categorising the electronic shop:

- According to the trading model (the one-to-one, one-to-many, many-to-many types described in the previous chapter).
- According to whether the market is buyer- or seller-driven (this is explained in detail later on in this chapter).
- According to scale (large or small).
- According to the kind of customer (retail or wholesale, for example).

We now describe a range of solutions for the presentation of goods that fit all of these. The practical step that needs to follow is the selection of the right solution for a specific eBusiness' size, customer base and market model.

3.1 CATALOGUES

A catalogue is the electronic equivalent of a shop's shelves, goods, departments, etc. It is the online representation of what is 'for sale' (or more correctly, what is available for trading). Some vendors use a catalogue to simulate a real shop (with 'departments' devoted to particular categories of goods and then 'shelves' containing the goods). Others structure their site more like a printed catalogue as supplied by mail order companies. However, in almost all cases, the trader will maintain some sort of catalogue application that is distinct from the details of the products themselves. In this section we look at these catalogue applications, how they work, the different types that are available and their strengths and weaknesses.

3.2 DIFFERENT SCALES OF CATALOGUE

Catalogues range in scale from a set of web pages and a simple script that allows orders to be taken through mid-range catalogue products that are characterised by a pre-defined structure of product categories and sub-categories, up to large scale corporate catalogues that are customisable and usually feature back-end integration with inventory, stock control and ordering systems. To a large extent, these different scales of catalogue reflect the kind of business that is being conducted online. For instance at the low end, a site such as http://www.norwest.currantbun.com offers only a few specialist products and services. Data is entered by hand and stored on a few backup disks. At the other extreme, an example is Maplin Electronics, a vendor of electronic components in the UK, which offers nearly 40,000 different products (see http://www.maplin.co.uk). Maintenance of the catalogue data (i.e. entry, formatting and quality of the data) is a job that demands the attention of a specialist. One well-known and very visible example of a high-end catalogue is Amazon.com, the online booksellers.

3.3 LOW-END DIY CATALOGUES

At the low-end of the scale, it is perfectly feasible to set up a simple retailing environment from a few HTML pages and an order form that generates an email message to the merchant. One step on from this is the putting together of a few scripts in Perl, Visual Basic, or some similar language to implement a simple 'shopping cart' model for buying items from the shop.

Such approaches are amply covered in existing books on HTML and Perl, and will not be dealt with in detail, here. However, it is worth noting the *limitations* of such approaches, of which the main ones are:

- It is impractical to build a shop that contains large numbers of products because the pages that describe them must be constructed entirely by hand. The Maplin Electronics site, for example, would be totally inappropriate to this approach.

- Changes to products are complex to implement. In some product areas, models and prices are continually changing, and it is desirable to be able to automate the updating of the catalogue.

- As with any site that includes large amounts of HTML pages and/or scripts, the maintenance becomes a major overhead (West & Norris 1997).

- There is no simple way to integrate with other systems including, payments, logistics and stock-control. Ideally, for example, the customer should be able to see whether items are in stock before placing an order.

For all these limitations this approach is commonly used, as it can get an electronic shop up and running quite quickly. The appeal of struggling with languages such as Perl (which is, most definitely, the province of computer buffs) is waning as cheap and easy-to-drive products such as Multiactive's EC Builder have appeared. These produce the same end effect with minimal pain. Indeed, commercial eBusiness products now exist that get your electronic shop up and running even if you don't have an established web site.

3.4 LOW-END COMMERCIAL CATALOGUES

At the time of writing, the signs are that low-end catalogues are likely to be bundled with basic web server software (e.g. by Microsoft) in the very near future, making it a very straightforward matter to build small-scale online shops, using 'out-of-the-box' components. The main issue for the vendor is then one of managing their own web site, and the main cost is probably that of the necessary permanent connection to the Internet.

Alternatively, a number of companies offer hosted catalogue sites at which the potential merchant can create an online shop without the overhead of running their own web site. A good example is the site operated by iCAT (a company now owned by Intel) at www.icat.com which hosts multiple 'virtual' catalogues for merchants. This architecture is illustrated in Figure 3.1.

In this model, the hosted catalogue site maintains multiple 'virtual' catalogues on behalf of numerous merchants. The merchants can use an online browser-based front-end to initiate their virtual catalogue and to manage its contents in-life. These tools are usually simple to use, but do necessitate a manual update process, so there is no opportunity for integration with the vendor's own systems, such as stock control.

Customers also access the catalogue over the Internet but, to them, each merchant site is quite distinct—they will probably be unaware that the particular merchant is sharing infrastructure with many others. This will

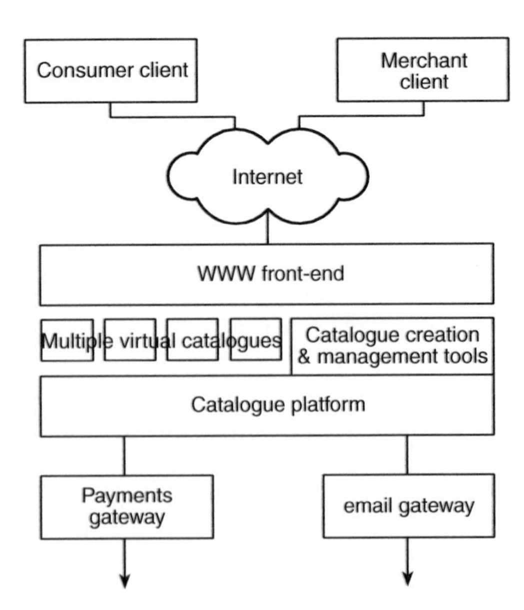

Figure 3.1 A virtual catalogue architecture

particularly be the case if the hosting service offers the vendor domain names which reflect the name of the vendor rather than the name of the hosting service.

By providing a fully managed service, companies such as iCAT address a range of major concerns that would otherwise have to be covered by the merchant:

- running their own web server and its associated computer hardware
- maintaining a permanent data connection to the Internet
- managing security components and processes: maintaining a firewall, monitoring the system for attempted attacks, etc.
- provision of standby servers to provide resilience in the event of component failure
- backup and archiving of data.

The hosted catalogue approach not only relieves the merchant of the need to maintain a web site, with all the attendant concerns listed above, but also allows merchants to share facilities such as payments gateways. Again, taking iCAT as an example, they provide access to a credit card gateway (for merchants who already have a merchant account with an acquirer), and will also email details of each transaction to the merchant's email address.

Customisation tools permit the merchant to provide a reasonable degree of customisation whilst taking advantage of the inherent structure of the catalogue software.

Overall, this approach is particularly suited to merchants who have an existing 'conventional' business and need a low-risk entry into online trading: it enables them to create a professional-looking online shop, containing a modest number of products, without making a big investment in in-house IT components and skills.

3.5 HIGH-END CATALOGUES

Whilst the types of catalogue described in the previous section offer many advantages, they do not satisfy all requirements. In particular, there are some specific high-end requirements such as:

- very large numbers of stock items
- highly customised sites which present a very distinctive brand image
- integration with other business processes such as stock control.

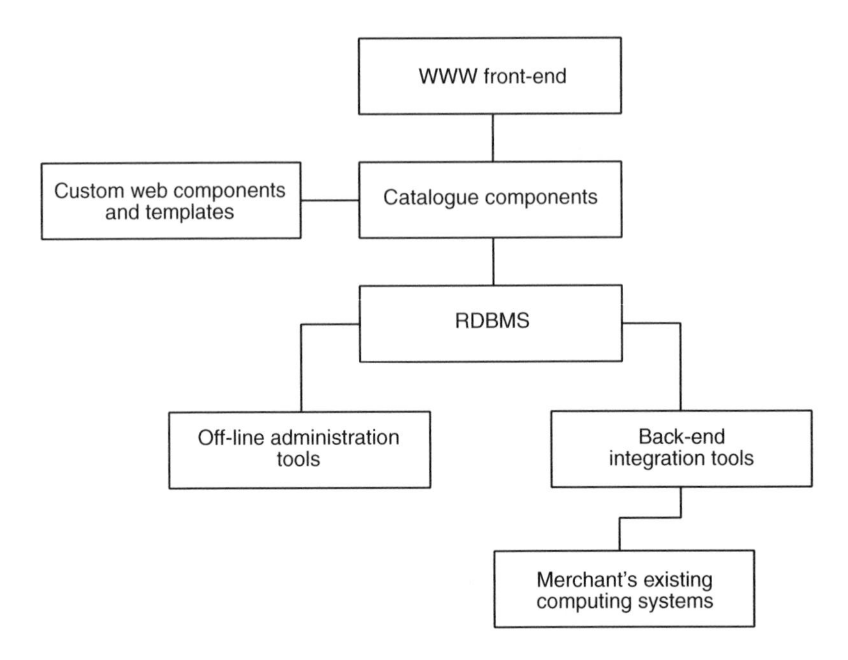

Figure 3.2 Structure for high-end catalogues

In the case of high-end catalogues, the typical architecture is illustrated in Figure 3.2. High-end catalogues are typically built around an 'industrial strength' relational DataBase Management System (RDBMS) such as Oracle or Microsoft's SQLserver. This approach has a number of benefits:

- RDBMSs are highly scalable, both in terms of the very large amounts of data that they can handle and in terms of processing large numbers of simultaneous accesses. In the latter case, such RDBMS products are typically capable of being scaled relatively easily from quite modest hardware, up to very high-performance multiple-processor systems.
- RDBMSs provide their own tools for routine operations such as data backup and archiving. They are resilient and can built to survive data corruption, disk failure, etc. Also, they usually come with flexible report generation tools. All-in-all they provide the essential but unglamorous tools that are not the main focus of catalogue vendors.
- RDBMSs provide standard interfaces for integration with other business applications, for example, using SQL, LDAP, CORBA, and other integrating protocols and technologies.

3.6 BUYERS AND SELLERS

Catalogues for buyers and sellers are different—the former is a virtual catalogue through which the buyer can see competing products from a number of suppliers, and the latter is a structured set of information that represents what a supplier has to sell. The technology used to represent the catalogue has to match one of these (i.e. optimised for one seller and multiple buyer, or *vice versa*).

So far we have concentrated on seller catalogues, but an example of a buyer catalogue is the Buynet product from BT, which aimed at companies who buy from a range of suppliers. A good example is that of supermarket chains who buy from numerous small suppliers: it makes sense for the

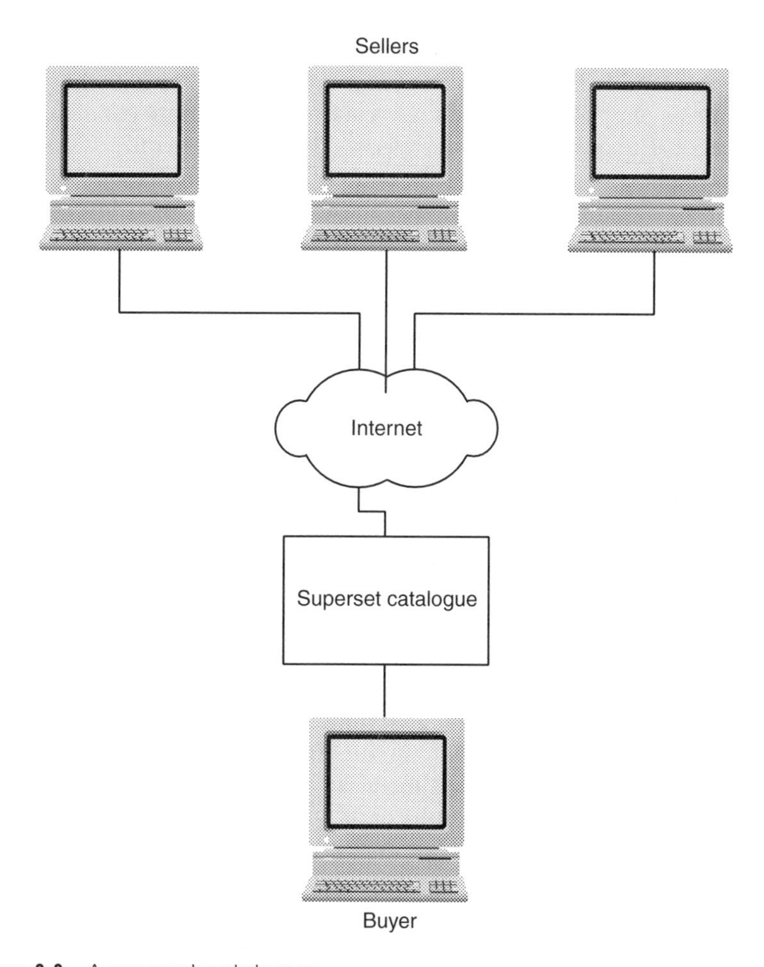

Figure 3.3 A superset catalogue

buyers employed by such a company to select from a single 'superset catalogue' that includes all the goods on offer from a range of suppliers, rather than having to search through numerous individual suppliers' catalogues. This architecture is illustrated in figure 3.3. Here, multiple sellers maintain the details of their own products in a 'superset catalogue' which can be viewed by the buyer. The sellers only have access to their own parts of the catalogue, whereas the seller is able to see the whole thing, including similar products that may be offered from more than one seller.

3.7 RETAIL VERSUS WHOLESALE

One further differentiation in catalogue that should be made is business-to-business as opposed to business-to-consumer. The consumer-oriented catalogues tend to be stronger on presentation as they usually have to sell on the basis of appeal. The business catalogues are more concerned with quick access to what someone needs. They are generally accessed by professional buyers who order the same items repeatedly, and may refer to items by obscure product codes rather than long text descriptions—pictures of products may be a total irrelevance (or at least an optional reserved for the occasional 'special' purchase). Similar distinctions exist in the 'real' world, where we do not expect the shopping experience at the trade counter of a car engine company to be the same as that in a high-street boutique.

3.8 MANY-TO-MANY MARKETS

So far in this chapter, we have only considered one-to-many market models (i.e. the various types of business-to-consumer retail catalogues) and many-to-one model (i.e. the buyer catalogue). The other case is that of many-to-many transactions. In the 'real' world, this is typified by the market concept of trading: a market included multiple traders who deal with a pool of multiple customers. In the online eBusiness world, many-to-many trading is usually associated with brokering or mediation functions. An example is insurance brokering, where multiple customers access the brokering site and their insurance requirements are matched against products offered by multiple insurance companies (Figure 3.4).

A number of current Internet sites offer insurance online, but many of these are not true online markets. For example, they may take the customer's details and email these to a 'physical' insurance broker, or they may simply match requirements against a few standard policies that are statically

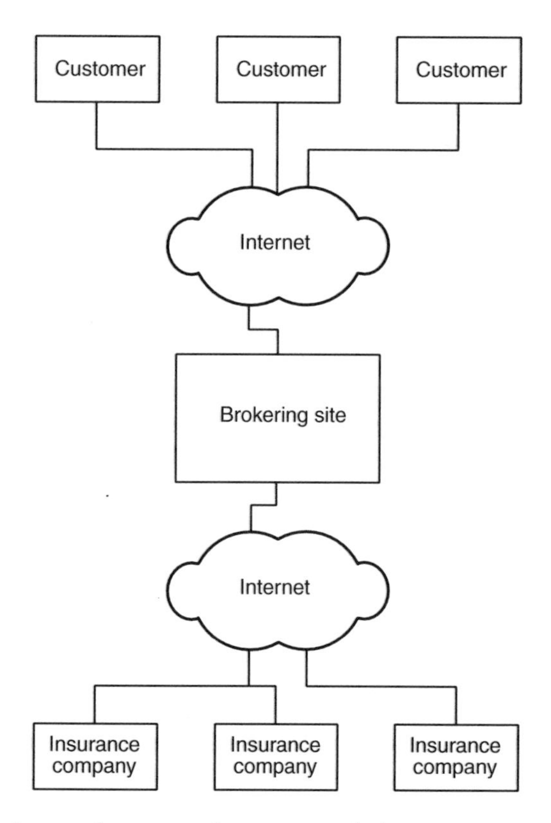

Figure 3.4 Brokerage in a many-to-many market

configured into the system. At the time of writing, there is, however, at least one system that is in trial (though not yet accessible to the public) which offers a true online market in insurance.

This system allows customers to buy insurance for their houses by accessing an online site. The way this works is as follows. Customers for house insurance first fill in a checklist which performs some pre-filtering, and directs them to a more specialised manual service if they fall outside the parameters of 'mainstream' insurance customers (e.g. if their house value is above a certain threshold; if they live in an area prone to flooding or subsidence, etc.). Those who fall in the mainstream category then fill in a more detailed online form which captures information about their property, the kind of insurance they require and their personal details. Participating insurance companies can input details of their policies, which must be describable in terms of the same parameters that are entered by customers (e.g. house value ranges), and also include the prices associated with these policies.

The insurance brokering site operates a 'quote engine', which performs a match of customer details with insurance policies and presents the customer with a set of alternatives (ranked in price order). The customer can then make an online selection and pay immediately by credit card. The details are forwarded to the insurance company, who then send the (paper) policy documents directly to the customer.

The brokering site handles the customer payment and takes a percentage commission on each transaction, before forwarding the balance to the insurance company.

3.9 MARKET MEDIATION

Closely related to the open market is the concept of market mediation: a service that, for example, acts on the purchaser's behalf to identify the best deal. We are all familiar with search engines which are the simplest incarnation of a market mediation product: the search engine trawls around the Internet to find information that most closely matches the consumer's requirements. The same principle can be applied to electronic shops, where a mediation service can, in principle, trawl round many merchants' sites for particular goods, and can hence find the best prices. The result is not a true open market, as the mediation service in this example is acting principally on behalf of the buyer. The example of the insurance brokering service given above is subtly different in that it is a true open market, where bids and offers are matched against one another.

Currently, the market mediation approach is somewhat limited, as there are no standards for online catalogues and these catalogues are, of course, aimed at human users rather than offering machine-readable output. The technology that everyone is looking at to solve this problem is XML—Extensible Markup Language.

XML is a markup language that has much in common with HTML, with which people are generally familiar as the language for creating web pages. However, HTML's markup tags only specify the *structure* of a page, i.e. they identify which part of a page should be treated as a heading, which part is a table, which is a footnote, and so on. Conversely, XML also allows for tags which identify items within the *content* of a page. So for example, a catalogue page could have XML tags which identify the name of a product, others which identify the price, and yet others which indicate the current number in stock. By agreeing such standards for XML tags amongst catalogue vendors, it becomes possible to create catalogue pages that can be read by machines as well as by humans. Once widely adopted, this will allow search-engines to 'understand' the nature of the content of a catalogue rather more effectively than at present, and therefore allow buyers to locate products (and suppliers) rather more easily than at present. Once search engines are able to provide more meaningful product searches they will start to become the 'front-door' through which a buyer can begin his search for suppliers. XML is explained in considerable detail in Chapter 7.

However, this vision of interworking catalogues is not without its practical problems. At the time of writing, there appears to be a risk of proliferation of Document Type Definitions (DTD) specific to identical business processes, and the random selection of tag names without guaranteeing uniqueness. This uncontrolled proliferation leads to lack of interoperability when it comes to data integration within applications. Companies such as Ariba and CommerceOne are already putting out competing DTDs. Some vendors envisage importing DTDs into applications with mapping facilities to map XML tag names to the application field names or provide 'data integration' tools that ease the integration into commercial applications—both these solutions are expensive and will therefore NOT appeal to the SME market. Hence, there is a pressing need for international standardisation work on XML standards for eBusiness transactions.

3.10 FULFILMENT

When we make a purchase in a conventional shop, fulfilment is often immediate, and we leave the shop with the goods in-hand (or more often in a bag with the shop's logo on it). In the case of the electronic shop, things are not so simple, as there must usually be some integration of the shop processes with some external fulfilment specialist such as the Post Office, Federal Express, etc. We consider this in more detail in the chapter on the supply chain, but there is one important case that belongs here, as part of the electronic shop: the direct fulfilment of 'soft' goods.

In this context 'soft' goods are items such as computer programmes, electronic documents, digitised music, and digitised pictures, all of which can be offered for sale, and can be delivered electronically over the Internet. Whilst it is simple to offer free downloads of files over the Internet, it is a little more complex to sell products which are then downloaded. Some of the considerations are:

- how to ensure that a payment has taken place before a download is permitted;
- how to cope with failed downloads due, for example, to network problems;
- how to protect the downloaded product from being copied by purchasers;
- how to deal with unsatisfied customers and returned goods.

Some examples of how to deal with these issues are as follows:

(a) *Free download of an encrypted file.* In this case, a simple file download facility is provided, but the file itself is encrypted and cannot be used by the consumer unless they have the necessary key. Buying the product is then simply a matter of buying the key to unlock the encrypted file. Typically, with this approach, the order might be handled manually and the encryption key distributed by simple email. This addresses the problem of ensuring that people only have access to the encrypted file if they have purchased the key, and it also addresses the problem of failed downloads: the consumer can make as many attempts as necessary to download the file, with delivery of the key being a completely separate process. The weakness of this approach is that for any simple implementation, the same decoding key is provided to all consumers, hence there is a risk of the key being distributed in an unauthorised way amongst potential consumers (e.g. by publication on 'hackers' websites). Nevertheless, this approach is simple to implement, and can use almost any encryption product for which files are self-extracting (i.e. no special application is needed by the person doing the decoding).

(b) *Demonstration versions with passwords.* As a further refinement of (a), some goods (particularly software) lend themselves to distribution in 'try-before-buy' demonstration form. The software may be time limited or may have some features disabled, but can be trialled at no cost. To activate the full version of the software, a key must be entered which 'unlocks' the full functionality. As with (a), this key can be ordered online and delivered by email. The product, in this case, is not actually encrypted, but must include its own internal policing of key entry. The main advantage of this approach is that it does allow consumers to trial the product before they pay, hence reducing the problem of 'returned goods'. Nevertheless. It still suffers from the problem of illicit distribution of keys. Also, it does not lend itself to purely information products for which 'try-before-buy' is not a meaningful concept.

(c) *Specialised soft-goods delivery platforms.* Rather than simply allowing the download of products and then providing a separate means of securing their content, an alternative is to use a more specialised commerce platform which monitors the process steps being followed by a particular user (including, particularly, the payment part of the process), and controls the ability to download the product, based on the steps that have taken place. Such systems generally require some form of user registration and log-on so that the platform can keep track of the activities of the individual. Download of the product is then mediated by an application. An example of this type of system is BT's Array trial service. Although this approach provides a lot of control, it still does not prevent the consumer from making illicit copies of the product and redistributing them (though this may be a lot more difficult than simply making illicit copies of keys. To overcome problems of failed downloads (e.g. due to communications problems) such systems generally allow the user to perform several download attempts within a preset time period (e.g. 24 hours).

(d) *Secure delivery infrastructures such as Intertrust.* To provide full control over material that could potentially be copied by the consumer, it is necessary to employ an infrastructure that includes some client-side components. For example, the Intertrust product includes a client which can integrate with various document viewers and common applications such as Microsoft Word. When products are is delivered to the consumer they include two parts: the information content in an encrypted form that can only be decoded by the specialised client and a set of rules (also encoded) describing how the content may be used. The client communicates with a network-based server, not only for the download process, but also at other points during the life of the product. The product itself cannot be viewed except under the control of the specialised client, and then only according to the rules that were encapsulated with it. With this level of sophistication, it is possible to set up very flexible options, for example: time-expiring content (it can be viewed for a set period, but then becomes unviewable unless a payment is made); pay-per-view (a payment must be made every time it is viewed); pay once on download and pay again for every copy made; and so on. This approach can solve all the major problems associated with soft goods delivery (except, perhaps, the most determined hacker) but, of course, comes at a price: in particular, the consumer must be prepared to run the special client on his or her PC, and accept that it will be communicating user actions to a network-based server.

3.11 OFF-THE-SHELF COMMERCE PLATFORMS

Increasingly, platform vendors are supplying out-of-the box commerce solutions which offer not just a catalogue product, but a whole set of tools for creating the electronic shop, including, payments, process flow and back-end integration. As an example of these, we look at the Microsoft Commerce Server, which is widely used by other vendors as the framework for creating commerce-enabled sites and applications. Microsoft Commerce Server is not just a shrink-wrapped commerce site: it also provides a set of tools for building and augmenting eCommerce capabilities. Some of the main components are:

- *Open Payment Architecture.* This a software architecture for payments, devised by Microsoft, which allows third-parties to develop components (called 'snap-ins') to handle any kind of payment processing. Microsoft's payment architecture not only supports multiple payment methods, but multiple secure payment protocols as well, including SSL and Secure Electronic Transaction (SET). SET, a standard driven by a number of leading industry companies including Visa and Mastercard, is a powerful secure payment alternative. SET is a three-way protocol, and manages the interfacing of consumer, merchant and financial institution in one single message. A number of independent payment software companies are delivering SET payment solutions based on the Microsoft Wallet and Site Server, Enterprise Edition API.

- *The Microsoft Wallet*. This is an ActiveX control or a Netscape plug-in that encapsulates and saves customer information (such as name, address, credit card number), so that the user doesn't have to type it in every time, hence simplifying the purchasing process. The Wallet also transmits the information to the server in a secure way. The information that is transmitted looks just like a standard HTTP post, so any server can read the information, not just Site Server Commerce. It is freely available on the Web, as well as being bundled with Internet Explorer 4.0, Windows 98 and Windows NT 5.0, and is already in use by Site Server Commerce and other systems like Intershop and Mercantec.
- *Order Pipeline*. Pipelines are Microsoft's way of modelling a business process such as a product order. The pipeline provides a framework consisting of a number of stages, each representing a discrete operation on a business object (e.g. an order form). At each stage, one or more specialised components operate on the object, then pass it to the next stage in the pipeline.

Fixed pipelines existed in Site Server Commerce 2.0, but version 3.0 allows customers to build their own pipelines.

Site Server Commerce starts you out with 14 predefined stages for the order pipeline:

—Product information
—Merchant information
—Shopper information
—Order initialization
—Order check
—Item price
—Item price adjustment
—Order price adjustment
—Subtotal
—Shipping
—Handling
—Tax

—Total
—Inventory adjustment.

The pipeline architecture is extremely flexible. Out of the box, Site Server Commerce version 3.0 has an order pipeline initially configured with the above stages. However, merchants can add or remove any stage if necessary. For example, if they wanted to add the capability for adding a gift message to a purchase, the merchant could add another stage in the pipeline to check if there was a gift message, and if so, validate the message and add a charge for it.

Another facet of the flexibility is in the components that process each stage. Site Server Commerce is bundled with a set of default components. However, there are also over 50 third parties providing components for many of the stages. For example, instead of a shipping component that simply calculates the shipping charge as a percentage of the order total, you can have a much more complex component that offers multiple shipping methods (e.g. ground or air) and calculates exact shipping costs based on shipping option and weight of the products.

Furthermore, all of the interfaces used to create a component are documented, so merchants can easily write their own components to customise their order pipeline even further.

- *Order Form Architecture.* An order form is an example of a business object, on which the components of a pipeline operate. The order form in Site Server Commerce is simply an object that contains key-value pairs. The values can, in turn, point to lists of more key-value pairs.
- *StoreBuilder Wizard.* This is a tool for simplified and managing online commerce sites.

3.12 AGENTS

One of the commonly-stated benefits of doing business on the Internet is the trend towards disintermediation—the removal of unnecessary links in the chain from raw product to end purchaser: the professional middlemen. The Internet makes it possible for the manufacturer to create a global sales channel direct to end users, without the need to set up physical shops and retail chains.

However, a contrary trend is also in progress which is opening up opportunities for new kinds of intermediaries, mainly based on the use of 'agent' technologies. A software agent has a number of properties: it has a degree of autonomy; it can learn; and it can interact with other entities such as other agents or with human users.

Some of the early systems that exhibit agent characteristics are Internet search engines that trawl the Internet on behalf of users, looking for particular information. The next step (and one which takes us straight into the realms of eBusiness) is the development of shopping assistants and shopping robots (also known as 'shopping bots'). A good example of these is the Inktomi shopping engine, which is employed by a number of shopping sites (see, for example, www.cnnfn.com). From the user's perspective, this site displays a list of categories of goods (electrical, clothing, sports equipment, photographic, etc.), and has a simple search field into which the user can enter a search string for the required product. So, for example, a customer who was interested in purchasing a digital camera would probably select the 'photographic' category and enter the words 'digital camera' as a search string. The shopping bot responds by searching Internet retail sites that are a likely source of products in this category (actually, the bot will have previously cached the contents of these sites and will only be performing the search within its own database, much like most Internet search engines).

The results that are presented to the customer are a list of products that match the description, together with an identification of the selling company, the price, availability, and buttons for links to give more detail, perhaps including detailed product specifications and consumer reports. The list can be ordered by price or other characteristics. If the consumer is happy with one of the offered products, he or she can press a 'buy' button, and be taken to a credit card entry screen to complete the transaction.

It should be emphasised that in this scenario the shopping site only hosts the 'bot' and its associated data cache. It is reliant on actual retailers around the net to provide details of goods for purchase. Hence, the shopping bot can be seen as acting on behalf of the consumer, enabling them to 'shop around' and identify the best price for their chosen product.

The trend seems to be towards combining such shopping engines with information-rich sites. In the real world a purchaser will perhaps buy specialist magazines (e.g. something like *What CD Player?* in the above example) and consumer publications, and will perhaps visit a number of stores to compare products and prices. In the Internet world, however, the same consumer may be able to access all this information from the single shopping site.

In the future, more complex agent interactions are expected to become commonplace. In particular, we are likely to see some in which several agents take part, not just the single agent that comprises a shopping bot. For example, a marketplace may involve multi-agent transactions where each agent takes a separate role: perhaps one representing a buyer, another representing a seller, and a third representing a broker. Each of these will be programmed to aim for its own set of goals, to best act on behalf of its human counterpart. For example, the seller agent might seek to maximise profit, whilst the buyer agent seeks to minimise cost and the broker maximises revenue.

Another good example of an opportunity to exploit agent technology arises in the travel industry (where the concepts of agents, albeit real live ones, is already well established!). Suppliers of various travel services might offer such services (flight bookings, hotel reservations, and the like) to the global Internet community via 'service agents'. A purchaser of travel services would not interact with these directly, but would employ the services of a travel 'broker agent', which would act much like a conventional travel agent. The broker agent would be able to take the customer's overall travel requirements and decompose these into individual elements (flights, hotels, car hire, and so on), and would interact with the community of service agents to obtain prices and availability for these elements, either individually, or in combination. For example, some service agents might be specialised for airline bookings, whereas others might offer whole ranges of travel services. The broker agent would then be able to offer the customer a set of priced options and complete the transaction online. Once again, there are extensive opportunities for 'information-richness' to add value to such an online service. For instance, the site could include video clips of travel destinations, virtual reality tours of hotels, or detailed itinerary maps.

Despite all this speculation, agent technologies are still in their infancy, and have yet to prove their worth to the hard-nosed business community. Although there are a number of good travel agencies online, they currently only sell packaged offerings or require the purchaser to decompose his travel requirements and enter them separately. Systems such as the example iven above currently only exist in demonstration form but, given the usual experience of developing products in 'Internet time', it is likely that commercial services will be online very soon.

3.13 SUMMARY

When you walk down any high street, there is an obvious difference between the shops you see. Some, like the banks, don't have any goods on display but just display lots of logos. Others have attractive displays with discrete price tags (if any), and others still pile their goods high and promise that you can't buy cheaper anywhere else. Each style appeals to a different part of us—sometimes we are driven by quality, sometimes by desire and sometimes by economy.

The subtlety of the high street needs to be recreated for the online shop, and that is where this chapter has focused. The various forms of catalogue—the statement of what is for sale—have been explained and their use discussed. It is the form of these catalogues that determines their purpose, with some optimised for volume trade at best price, some for specialist purchases. Hence we have illustrated how the various types are built, and how they can be deployed to good effect.

REFERENCE

West S & Norris M, *Media Engineering*, John Wiley & Sons, 1997.

4

Payments, Credit and Invoicing

If you want to know what God thinks of money, just look at the people he gave it to

Dorothy Parker

Business only exists so that people can improve their lot in life. This might involve an exchange of knowledge, tokens or favours, but the usual currency of business is hard cash. Given the inherent fluidity and transience of the virtual market, the transfer of real money is not an easy thing. Sellers want assurance that they will be paid; buyers don't want to give any more than is due (and certainly not private details of their bank account), and both parties want a secure, reliable and usable mechanism of honouring their commitments.

Progress in the technology behind electronic payments systems has been slow but steady over the last few years. The main reason for this apparent lack of urgency in a central part of eBusiness is that effort has been focused on integrating the various components needed in the complete purchasing process to produce end-to-end solutions, rather than in the actual payment systems themselves. So, despite much hype, online payment of items is still not as widespread as predicted, with a lack of trust of the payment method cited as one of the top reasons why users don't buy online.

This chapter deals with both the technology of on-line payments and the mechanisms for establishing trust in that technology. Before we go into either of these, though we need to be clear in what we are talking about.

4.1 SOME DEFINITIONS

It is said that tyranny begins with a corruption of language. The diversity of terminology associated with electronic business certainly shows all the signs of attempted dictatorship (mostly from the technical community). So, before we start explaining the mechanics of on-line cash, a few key definitions:

- *Invoicing.* This is the first step in the payment process, and is usually associated with electronic invoices, either through email or through a system that provide an online view of an account.
- *Clearance.* The next step is the transmission, reconciliation and, in some cases, confirmation of the payment orders that matches the invoice before settlement takes place.
- *Settlement.* The process of recording the debit and credit positions of the parties involved in a transfer of funds. Settlements can be 'gross' or 'net'. In a gross settlement, each transaction is settled individually. In a Net settlement, parties exchanging payments will offset mutual obligations to deliver identical items (e.g. UK pounds or US dollars) at a set time, such as the end of the day. At this point, only one net amount of each item is exchanged. For example, if I owe you £10 plus $3 and you owe me £6 and $11, we would probably conduct a net settlement in which I'd give you £4 (10 minus 6) and you gave me $8 (11 minus 3).
- *Collections.* This is the last part of the payment process, when one account is credited, another debited. A collection is assumed to be completed once authorisation from the financial service provider (i.e. bank) for the charge is confirmed. Collection is not that straightforward with, for instance, recurring charges with credit cards not allowed under UK SWITCH rules. This, and Cardholder Not Present rules resulting in possible chargebacks, create a requirement for debt management facilities.

4.2 PAYMENT TRENDS

Attitudes to the use of different payments mechanisms are changing (and vary when considering Europe, US or the whole world). The major options for paying for goods and services include:

- cash
- cheque

- direct debit
- credit card
- debit card
- credit accounts
- others such as 'barter', which may not be taken seriously in international commerce but which have a potential counterpart in eBusiness. They will, therefore, not be completely neglected in this chapter.

The preference for one or other of these mechanisms varies according to whether the transaction is business-to-business (for which a credit account is common), business-to-consumer (for which credit card transactions are favoured) or individual-to-individual (in which case cash or cheque are most common). Within the business community itself, there are also variations, with the smallest businesses being more likely to use credit cards for their purchasing. Also, some industries are particularly averse to credit whilst others accept it as the norm, and it is noticeable that some parts of the world and some cultures are more conservative in their approach to payment options than others.

Table 4.1 is based on an Association for Payment Clearing Services (APACS) report of July 1998, predicting expected trends over the period 1997–2007. It shows the current and expected values of UK transactions (in pounds sterling) for the major payments methods.

Table 4.1 Predicted and current values of UK transactions for major payment methods

Payment method	Volumes—1997	Volumes—2007	% change on 1997
Cash	25.5 bn incl. 14bn < £1	Not available 1.4bn e-cash	
Cheques	2.8bn	2bn	−28%
Credit Cards	1.06bn	2bn	+88%
Debit Cards	1.5bn	3bn	+100%
Automated			
Direct Debit	1.6bn	2.9bn	+81%
Direct Credit	0.9bn	1.5bn	+66%
Standing Ord	0.3bn	0.3bn	No change
Total Automated	2.8bn	4.7bn	+67%

Note: the figure for e-cash in 2007 assumes a 10% market share of low value transactions (newsagents, vending, ticketing, etc.) by that date. Cash as a percentage of all transactions is forecast to fall from 75% to 63% between 1997 and 2007.

4.3 MAIN INDUSTRY PLAYERS

When considering the likely evolution of the online payments industry, it is essential to take account of the major players involved in 'conventional' payments systems. The main players in the payments industry are:

- *Global payment service providers such as Visa and Mastercard.* These few companies have a massive global presence in payments services. For example, Barclaycard operates in 247 countries, has over 13 million merchants and over 650 million cards. The challenge for them is to assure themselves that there are adequate security processes to support online variants of their traditional processes. To-date, these have been a major force in the promotion of online payments solutions.
- *National payment associations such APACS and LINK.* These organisations have been established to fulfil specific branding, standards or processing needs of their member banks. The payments associations in the UK and Europe are generally specific in terms of their individual responsibilities and roles. For example, the Association for Payment Clearing Services (APACS) is the UK Membership standards body, with 23 member banks which oversees money transmission in the UK and has the responsibility for the co-operative aspects of payments including plastic cards. Similarly, the Bank Automated Clearing Services (BACS) is responsible for bulk electronic clearing, and handles automated payments such as Direct Debits, Credits and Standing Orders in the UK.
- *Card Issuing Organisations such as Banks and retailers* (e.g. Barclaycard in the UK).
- *Merchant acquirers* (such as Barclays, HSBC and NatWest in the UK).
- *Terminal/device/card manufacturers* (Racal, Verifone, Delarue).
- *Third party processors and personalisation bureaux* (FDR, CCN, Orga).
- *Telecom and network providers* (BT, AT&T, etc.).
- *Systems integration and security specialists* (ICL, IBM and EDS).
- *Standards bodies and industry consortia* (IETF, W3C—more on these later).

The other consideration is companies who are dominant in related industries but could rapidly move into payments. Of these, the most significant is Microsoft. By owning the standard for computing client software environments, they can potentially move very rapidly into the client end of electronic transactions (for example, they could bundle an electronic wallet with a future version of the Windows operating system). In October 1998, Microsoft announced the 'Smart Card for Windows' operating system, a move expected to provide strong competition to the current industry leader MULTOS (see the Mondex section below). At the time of writing, there are also strong rumours of Microsoft entering the banking market.

4.4 CREDIT AND DEBIT CARD PAYMENTS

In looking at electronic payment systems, we take each of the main 'conventional' payments approaches, in turn, and consider how it is evolving to support eBusiness.

Credit and debit cards are considered together as they are normally processed in the same way. These have been the main focus for early implementers of online payments systems. The roles in a credit card transaction, whether electronic or otherwise, are as illustrated in Figure 4.1.

In the scenario in Figure 4.1, the consumer wishes to buy goods from a merchant and to pay with a credit card. The consumer must first open a credit card account with an 'issuing bank' who provide them with the card itself. The merchant must also hold an account, in this case with an 'acquiring bank' who are prepared to process credit cards. The acquiring bank also provides the merchant with a means of recording transactions (e.g. an electronic point of sale device or a manual device for taking card

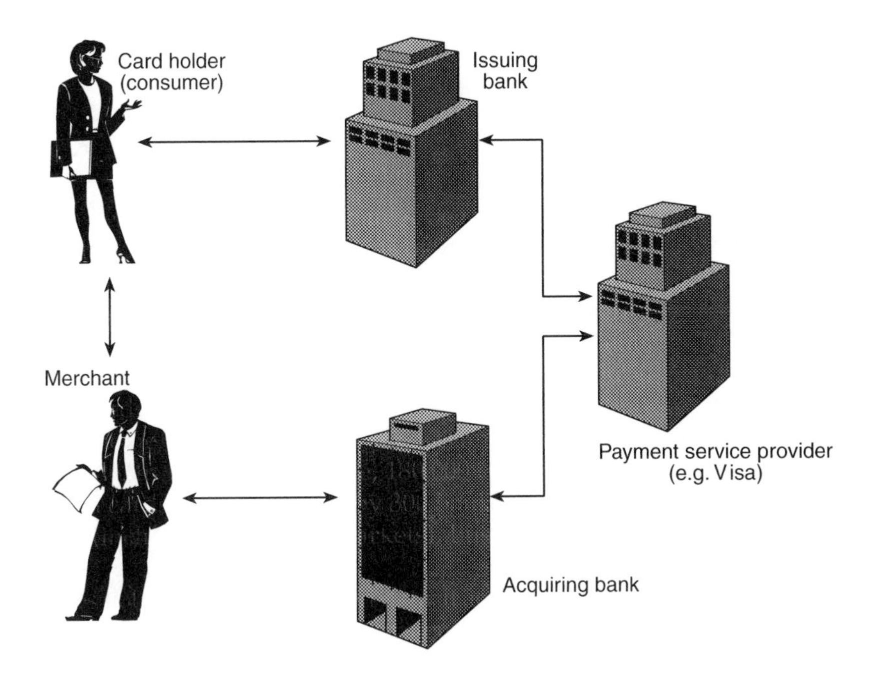

Figure 4.1 How an online card transaction works

imprints). The additional party in the transaction is the payment service provider (i.e. the credit card company, such as Visa or Mastercard).

The clearance and settlement of credit card transactions is a three-stage process:

- Authorisation is the process in which the issuing bank approves (or declines) a proposed transaction at the point of sale.
- Clearance is the process by which the acquiring bank collects data about a transaction from the merchant and delivers this data to the credit card company who, in turn, deliver it to the issuing bank who use this information to add the transaction to the cardholder's account.
- Settlement is the process by which a credit card company collects funds from the issuing bank and pays them into the merchant's.

In a typical sequence of transactions, the card holder may enter a merchant's shop and offer a credit card in exchange for goods. The merchant records details of the credit card and the transaction details, usually swiping the card through a Point Of Sale (POS) terminal.

These details are forwarded via the credit card company to the issuing bank, who then approve or decline the transaction based on the cardholder's account status. This approval or disapproval is transmitted electronically to the shop via the credit card acquiring bank. The network from merchant to acquiring bank is generally PSTN or the old-fashioned but reliable X.25 data network (e.g. BT Cardway service).

Clearing takes place, probably as an overnight function, where the merchant submits all of its credit card transaction data electronically (credit card drafts) to its acquiring bank. The acquiring bank then credits the merchant's account for its transactions, and the merchant is now out of the loop. The acquiring bank now needs to get paid for the transaction, so it sends the transaction data electronically to the credit card company who distribute it to the appropriate to each issuing bank.

The final step is settlement, in which the credit card company collects funds from the issuing bank's account and transfers them to the merchant's account with the acquiring bank.

After settlement, the issuing bank will present the transaction as an item on the cardholder's monthly statement, and once the cardholder pays the card-issuing bank, the cycle will be complete.

The link from the merchant to the acquiring bank is generally already electronic in nature (over a secure private network), as is the link between the banks. The focus of current eBusiness developments is in two areas:

- Allowing the transaction between the cardholder and the merchant to take place over the Internet.

- Automating the function of the merchant in processing the card and transaction details and interfacing them into the private network to the acquiring bank.

There are several challenges to achieving these aims. The main ones are as follows:

1. Cardholders perceive the Internet as inherently insecure, and will not send card details 'in the clear' over the public network.
2. The cardholder and merchant both need to trust that the other is who they purport to be.
3. Even with confidence that the merchant is 'genuine', cardholders are reluctant to give their card details to a merchant with whom they have had no face-to-face contact.
4. Acquiring banks are reluctant to accept responsibility for 'Cardholder Not Present' (CNP) transactions. Under UK law, the risk for these transactions is taken by the merchant.
5. Merchants must be able to cope with refunds to customers.

There are several practical strategies for handling credit card payments over the internet and these will be described in sequence of increasing complexity. We assume that a company wishes to sell goods to the public and to accept orders and credit card payments over the Internet.

Option 1: Insecure credit card details + manual processing

In this case, when the consumers select goods for purchase, they are presented with a web form into which they must enter their credit card details. On completion of the form, they typically press a 'Submit' button, which causes their web browser to send the details across the Internet to the merchant's web server, in a form that is processable by a CGI application. The CGI application processes this information and presents it for manual processing by the merchant. This can be done in a number of ways: the CGI application might email the merchant with the transaction details; it might simply store the details in a database which the merchant can access through a web form; or it might simply print out the transaction details in the merchant's premises, for manual processing (Figure 4.2).

Having extracted the payment details, the merchant types them into a POS terminal from where they are transferred across the usual banking network to the acquiring bank, which sends back an acknowledgement that the card is valid.

This approach has some advantages: it is simple to implement, and is particularly so when implemented as an add-on to an existing trading business. An existing retailer who already has a merchant account with

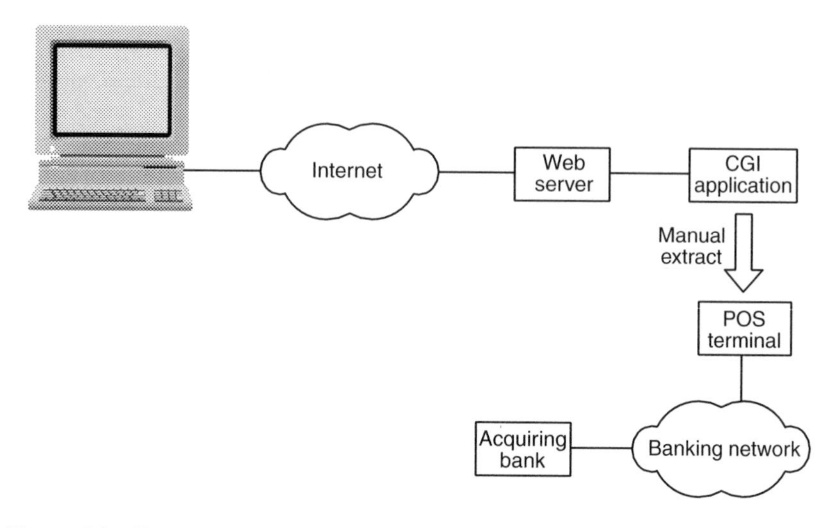

Figure 4.2 The simple strategy for credit card payment

an acquiring bank can easily set up a web site to take Internet orders alongside 'conventional' orders. Indeed, the merchant's infrastructure can be simplified even further if the web hosting and CGI application are undertaken by a third party 'Internet shopping bureau' who, for example, email or even fax the orders through to the merchant. iCat in the US and BT Webworld in the UK are examples of companies who offer this kind of bureau service.

The main disadvantages are in the area of security: there are no security features in the Internet access between the consumer and the merchant, and the merchant (as well as, perhaps, some third party bureau) is storing the consumer's credit card details. Such details could be misused by an unscrupulous merchant. From the merchant's point of view, although he may have had an acknowledgement from the acquiring bank that the credit card was valid, this is still a 'Cardholder Not Present' transaction, and the merchant must bear the risks such as that arising if the card (or even just its details) have been stolen by the consumer. A further disadvantage of this option is that there is no attempt to automate the functions of the merchant.

Option 2: Use of a server-side encryption key to secure the link

As described in the chapter on trust and security, privacy across the Internet is commonly achieved by the use of an encrypted link using, for example, the standard 'secure sockets' layer that is commonly used by web servers and browsers. The initial interaction between browser and server takes place using a form of public key cryptography where one key (the public key) is used to encrypt the message and a different key (the private key) is used to decrypt it. In the simplest set-up, described here, the private key is held at the server end, and matching public keys are made available to clients. This arrangement provides two benefits: first, the transfer of the consumer's credit card details takes place over an encrypted link, removing the most significant weakness of Option 1. Secondly, provided the merchant is distributing public keys that have been 'signed' by a trusted third party, the consumer can be confident that they are giving their card details to someone who is who they purport to be.

This level of security can be provided as a 'bolt-on' option to most web servers. A server-side certificate can be purchased from trusted third parties such as VeriSign in the US and TrustWise in the UK for a few hundred dollars.

However, this approach does nothing to automate the processes of the merchant and still allows card details to be misused by an unscrupulous merchant, even though they are not, now, accessible across the Internet.

Option 3: Use of an automated card gateway

Merchants who deal with large volumes of goods can benefit greatly from the automation of the process by which credit card details are extracted from the web server application and are entered into the acquiring bank's network. In other words, the manual extraction process is replaced by a payments gateway (Figure 4.3).

Examples of payments gateways include Soft-EFT from Commedia and PC-EFT from Servebase Computers Ltd (http://www.servebase.co.uk). Soft-EFT, for example, is a suite of programs that enable processing of payments by Visa, MasterCard, Switch, American Express, JCB, Diners Club and Style cards from a PC. The programs have been designed for quick and easy integration with merchants' applications.

The network described in the Figure 4.3 as a 'banking network' can in reality be one of a number of physical networks: most acquiring banks support PSTN, ISDN and X.25 access, for example. The messages that flow across this network are those defined by APACS (the Association for Payment Clearing Services), and are known as APACS 30 and APACS 50.

APACS 30 defines the method used for the electronic *approval* of credit card transactions, whilst APACS 50 defines the transfer of the details that make up the transaction itself. Acquiring banks define a 'floor limit' of transaction value, below which the merchant can authorise without a call to the bank by scanning a 'hot list' of stolen cards; transactions with values above the floor limit communicate with the acquiring bank using the APACS 30 protocol. This process is concerned purely with authorisation, and does not involve the transfer of funds. This mode of operation is described as 'predominantly offline'.

The APACS 50 standard defines a second set of protocols by which the

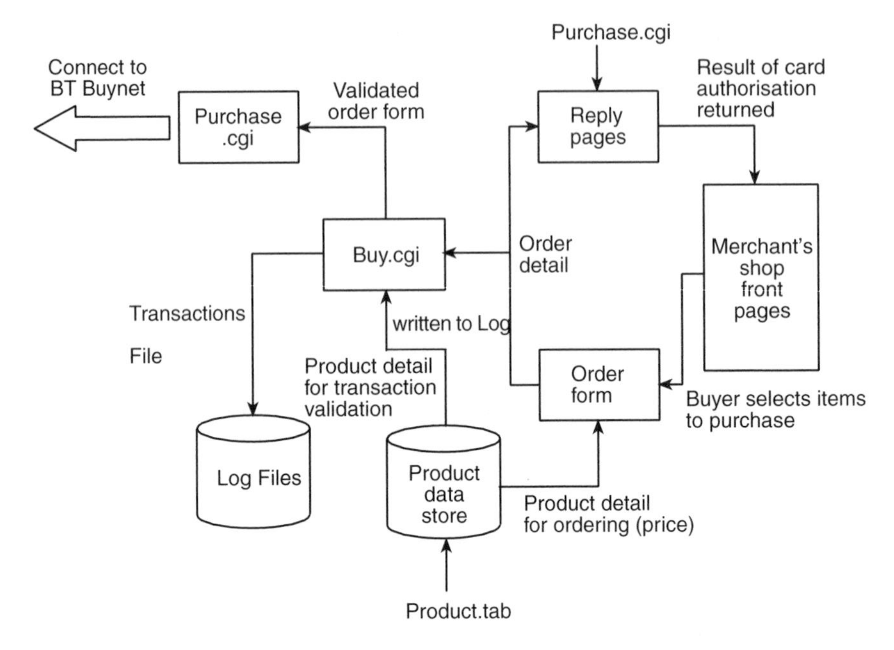

Figure 4.3 A more sophisticated strategy for credit card payment

acquiring bank makes a connection to the payments gateway, overnight, to retrieve transactions and update the hot list file. The information collected by the bank at this time is used to transfer funds within a few days.

The main issue that is not resolved in this option is that of the holding of consumers' credit card details.

There are really two distinct models for credit card gateways. The first one where the client asks for some goods, interacts with the merchant, provides payment details which the merchant passes on for settlement. In the straight through model, the merchant retains credit card details (for refunds) for some time.

The second is where the client requests goods from supplier who refers the buyer to the payment service (Clear Commerce, Open Market). In this second instance, credit card details do not go to the merchant at all—the 'buy' button resolves into a URL, and when the link is enacted, all details of the transaction are sent to this site (buyer, merchant, goods, cost).

Option 4: Use of SET standards

The ultimate vision for payments is a regime where there is privacy of information to ensure that only those parties who need access to it are able to obtain access, and in which all parties trust the identities of all other parties. Also, there is the need for individuals to be able to authorise

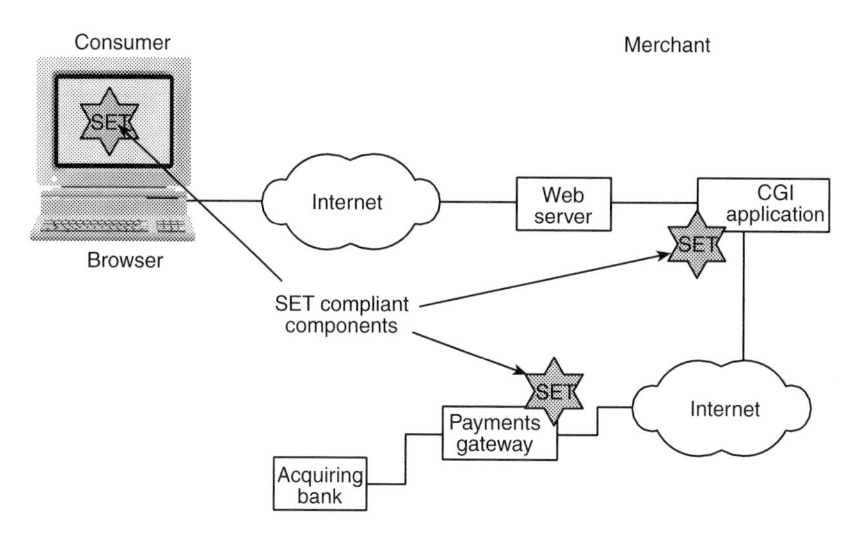

Figure 4.4 Secure Electronic Transaction in action

transactions (i.e. 'sign' them). All these concerns are addressed by the SET (Secure Electronic Transactions) standards.

In Figure 4.4, the payments gateway is shown at the acquirer end of the chain as there is no longer the need to use a private network from the merchant to the acquirer: the whole transaction can take place over the public Internet up to the point where the transaction details get decrypted (i.e. at the payments gateway), so it makes sense for this to be more closely associated with the acquirer.

Risks in credit card transactions

Major fraud risks in credit card transactions include:

- unauthorised use of lost or stolen cards. These account for some 50% of all credit card fraud)
- fraudulent applications for cards (e.g. using a bogus identity)
- forgery or alteration of cards
- fraudulent use of a cardholder's credit card number
- Consumer delinquency and default. If a cardholder fails to pay for the charges, then the issuing bank is liable to pay the acquiring bank.

The merchant is responsible for paying any costs related to credit card fraud if the merchant does not do at least one of the following three things: obtain an authorisation; the cardholder's signature; or the electronic imprint of the card.

Other technology trends to reduce credit card risks

- *Neural network systems.* These allow issuing banks to track the cardholder's spending patterns and to detect any spending discrepancies, thereby detecting potential credit card fraud. For example, if a cardholder, who typically only purchases small value groceries, suddenly starts buying large amounts of high value consumer electronics goods, the system can alert the issuing bank of potential fraud.
- *Address Verification Service (AVS).* AVS allows telephone order companies to verify a cardholder's billing address online. This program is designed to reduce fraudulent use of a cardholder's credit card number. Using AVS, the telephone order companies can check the delivery address provided by the customer against the billing address on file with the issuing bank. If the two addresses are different, suspicions are raised.

Issuer's Clearing House Service (ICS)

In the US, Visa and MasterCard have developed a clearing house database called Issuer's Clearing House Service (ICS), which is intended to detect fraudulent applications for credit cards. ICS allows issuing banks to compare application details with a database of invalid addresses and Social Security numbers. The ICS database includes information such as Social Security numbers, names and dates of birth of credit card applicants.

4.5 ELECTRONIC CASH

Despite the plethora of sophisticated payments instruments that have been dreamed up by financial institutions, there is still a widespread attraction to the use of cash. This is particularly the case in small-scale consumer transactions, but is also commonplace in some businesses. However, the cost of handling cash is high: IBM estimate that the cost to banks of handling cash transactions amounts to 30 billion Euros per annum worldwide, and a Boston Consulting Group study estimated that the costs to UK retailers and customers amounts to £4.5 Bn per annum. It may seem a contradiction in terms to speak of 'electronic cash', but it is clearly possibly to develop a payment component that has all the characteristics of cash (except, perhaps, the tactile ones associated with crisp paper money or chunky coins!) but is conducted purely electronically.

For example, a number of systems (of which the best known is Mondex) allow consumers to make electronic payments through the use of a so-called 'stored-value card'. These may physically resemble a credit card, but are different in several respects. First, unlike a credit card, a stored-value card digitally stores an actual monetary value (not just some account details); secondly, when a transaction is made, the monetary value is instantly removed from the card, like money from a physical wallet. Hence, users of stored-value cards do not necessarily need a bank account, and merchants need not verify the cardholder's identity whenever purchases are made. The market for stored-value cards is that of low-value transactions too small for economical use of a credit or debit card, and it therefore has some shared characteristics with micropayments systems (see Section 4.9). Indeed, a stored value card may be thought of as a pre-pay micropayments system which uses client-side storage.

Some stored-value cards are described as 'single-purpose' in that they can only be used in one payment scenario. Common examples of these are phone cards. Within the realm of 'general-purpose' cards that can be used to buy goods and services from different vendors, some are disposable: they are pre-loaded with a fixed monetary value that gets decremented every time the card is used until it is completely exhausted, at which point the card must be discarded. Others, however, are reloadable and can be replenished by inserting them into a device such as a specially equipped ATM machine or a telephone-based terminal. One objective of eBusiness is to allow people to load-up their stored-value cards using a PC connected to the Internet.

The economic viability of these cards remains unproven, to-date. One major obstacle is that for their use to reach a critical mass, a huge investment must be made in ATMs, cash registers and other devices that might be involved in crediting or debiting the cards.

Mondex

In July 1995, Mondex, a venture of two British banks, began conducting a test of stored-value cards with the 190,000 residents of Swindon, England. The cards and the merchant interfaces were provided free of charge. Swindon residents were offered reloadable cards to be used at a majority of the town's shops, pay phones, and buses. There were 8000 residents who actually used the cards. Of the 1000 merchants, in Swindon, 750 signed up to accept the cards.

Through a franchise scheme, Mondex have initiated 23 implementations of their system worldwide. In Hong Kong alone, 180,000 cards had been distributed by 1999, with cards being accepted by 300 bank branches, 700 ATMs, 7000 merchants (including 220 supermarkets), buses, swimming pools and others.

The Mondex card does not just contain encoded data: it actually contains a single chip computer running its own operating system called MULTOS. This operating system allows multiple applications, written by different companies, to be run (and to be securely stored on the card). Furthermore, applications do not have to be written into the card at manufacture: they can be loaded onto, or deleted from, the card, for example when it is inserted into a terminal. Currently, two Mondex/MULTOS chip implementations exist: one from Hitachi and the other from Siemens. The thing that sets Mondex apart is that it behaves exactly like real cash: for example, one individual can transfer 'cash' to another using a 'wallet' that allows the insertion of the two people's cards. Such a transaction can take place without the involvement of a third party (i.e. a bank).

BarclayCoin

The BarclayCoin system is offered by Barclays Bank in the UK. It consists of an application that can be downloaded and installed on a consumer's PC, which is then bound to an existing credit or debit card. Funds can be loaded into the wallet by debiting the card, and are then available for online shopping.

Visa cash trial (1996 Olympics)

At the 1996 Olympics in Atlanta, Visa and the three largest banks in the southern United States conducted the largest experiment with stored-value cards. About two million stored-value cards were made available in denominations of $10, $20, $50 and $100. All of the 85,000 spectators at the opening ceremonies were to be given $5 cards. In July 1996, 198,000 transactions were made with Visa cash cards.

Manhattan pilot

Chase Manhattan Bank, Citibank, MasterCard, Mondex and Visa announced an extensive pilot project on the upper west side of Manhattan to begin in October of 1997. Between 50,000 and 100,000 bank customers were offered reloadable cards that were subject to predetermined dollar limits. About 500 merchants agreed to accept the cards.

E-Cash

DigiCash was the first company to produce an Internet-based electronic money product, which it calls e-cash. Its creators claim that e-cash combines the speed of network-based payment systems with the anonymity of cash. As such, it is a little different from the card-based systems discussed so far. In the case of e-cash, the cash is stored on the user's PC.

E-cash is a digital representation of a monetary value which is stored on the customer's hard drive. The customer uses this digital representation to pay for the transactions that he or she makes. Amounts of money can be withdrawn from a bank account and converted into e-cash; e-cash can be debited from a consumer and credit to a merchant; and e-cash can be 'paid in' to a merchant's bank account.

4.6 CREDIT ACCOUNTS

Up to now we have been considering the cases where goods are purchased and a payment is made immediately (although the payment may be made from a credit card account which is not settled for some time). However, in many cases, 'conventional' transactions are not paid at the time the order is placed: rather, goods are purchased and the vendor records the details in a credit account. After a certain period of time (or at certain intervals), the vendor presents the purchaser with an invoice which must be paid within a certain period. This is the usual practice in business-to-business transactions and in certain business-to consumer transactions: for example, telephone and utility bills.

The payment part of this process is no different from any other that we have dealt with already (i.e. the invoice could be paid by credit card, direct debit, or even cash). However, if you are going to pay your bills online, it makes sense to receive them online too, so the main issue we are considering here is the online delivery of invoices. This is referred to as 'bill presentment' or 'bill presentation', and we shall adopt the latter term here.

Bill presentation

Electronic bill presentation not only offers a direct replacement for paper bills: it also opens up new possibilities such as being able to check one's bill at any time, not just receiving it at the end of a preset period. Hence, the main considerations here are:

- Online presentation of billing information.
- Integration into backend systems to offer additional billing options such as real-time viewing of bills.
- Meeting legal requirements including tax law.
- Associating the bill with the subsequent electronic payment.

The conventional billing process involves the stages, shown in Figure 4.5 and listed below:

- *Collection of billable events.* These could be anything: the number of

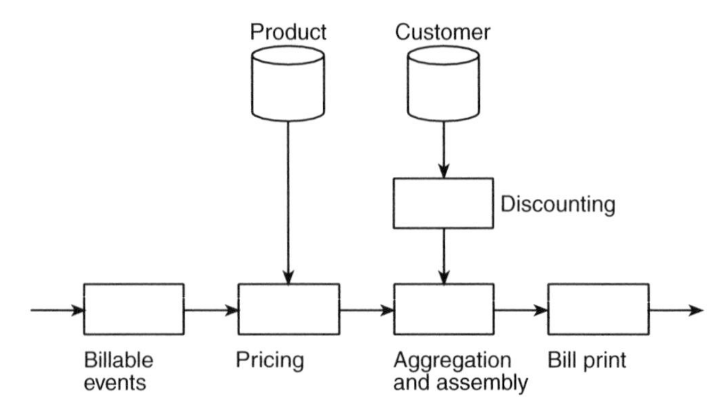

Figure 4.5 The traditional range of billing functions

seconds spent using a particular online service; the purchase of goods, etc. These have to be collected in real-time.

- *Pricing and charging.* This is derived from a product and service portfolio, which gives prices for each component and for packages (e.g. free web hosting if you purchase our settlement service).
- *Discounting.* The allowances given to a specific customer (usually on the basis of their size and purchasing power).
- *Aggregation and assembly.* The compilation of a bill for a specific customer based on all of the products and services they have, customer and product discounts. Usually straightforward for an individual, but can be very complex for a large company.
- *Bill printing.* This is normally a batch process, run at an appropriate time to generate the bill.

In eBusiness, it is this final stage that must be intercepted to generate an electronic bill. In the simplest case, this might simply be a replacement batch process which generates emails containing the billing information. However, by intercepting the process at an earlier stage, it is possible to offer an online billing service that is closer to the real-time ideal.

The Internet opens up possibilities not only of seeing your bills online, but also the creation of a commonly accessible clearing house for bills, or billing bureau service. In these, a customer would have a single account in which they receive bills from numerous sources (including utility companies, credit card companies, store cards, etc.). The bureau provides a single point of reference for everything you owe. It also lets you look at billing history, so you can check for anomalies.

An outline of a billing bureau is illustrated in figure 4.6. In the diagram, the service would be paid for by the large companies (Steve-Mart, Groceries-R-Us, etc.) shown on the left-hand side. Their motivation is clear (and worth

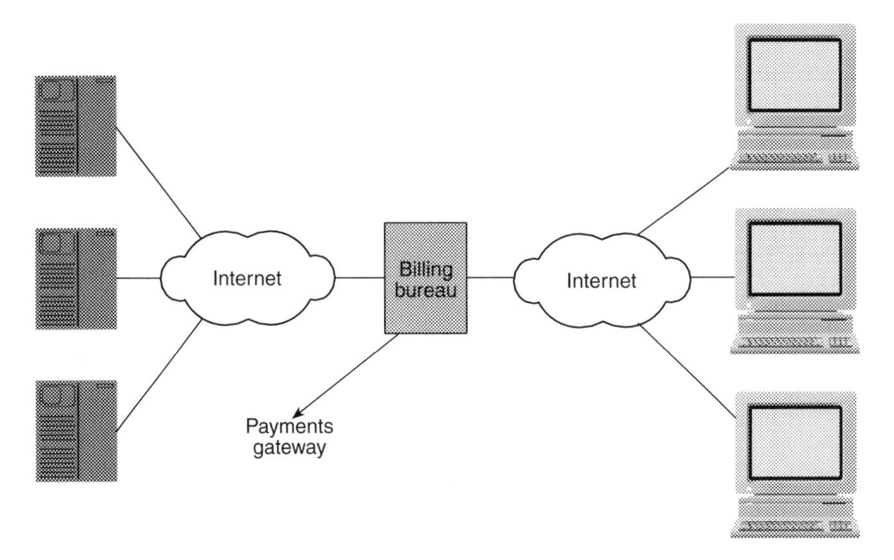

Figure 4.6 A billing bureau

spelling out); as large buyers, they save money by trading electronically, and their suppliers are motivated by the fact that they can only get onto the qualified supplier list of major corporates if they are online. Hence, the buyers use their purchasing power as a lever to push eBusiness forward.

The bureau includes a separate account for each consumer. Bills from multiple sources (utility companies, stores, etc.) are delivered to the bureau. The consumer is then able to browse these, at leisure, in a single location. A 'pay' button would allow the bill to be paid. The 'single inbox' for bills can be customised so that, for example, an SMS message might be sent whenever a new bill arrived. It would also allow people to compare current bills with a previous billing history

For most individuals, the billing bureau service would be a welcome change for reams of paper that always seem to get lost when you need them. The prospect of having configurable email or Short Message Service (SMS) reminders would help many to avoid late payment surcharges!

For companies generating the bills, there is an obvious cost saving in printing and postage (and this runs into many millions of pounds for some companies). In addition, there is the indirect benefit of creditors being better organised.

Bill payment

Presenting the bill is only one part of the picture: the other is paying it! Electronic bill payment services are offered by a number of banks and other institutions who will (for a fee) pay designated bills, after authorisation, on a consumer's behalf. Such payments may be made electronically or by printed paper cheque. If the payee does not accept electronic payment, the bill-paying service would print and mail a cheque on behalf of the consumer. This type of payment system is not very different from a customer writing cheques: the only difference is that a centralised computer system is involved in delivering the cheque to the payee.

Some banks provide bill payment services to their customers, or may contract with an outside electronic bill payment service to provide the service for the bank's customers. In addition, consumers may contract independently with an electronic bill payment service; in such a case, no contractual relationship exists between the bank and the service.

To communicate with the bill payment system, the consumer uses a home computer equipped with a modem and appropriate software. Messages are transmitted over a private communication network (not the Internet).

Electronic bill payment providers recommend that consumers schedule payments three to four days before they are due, in case the payment has to be sent through the mail instead of electronically, and also to allow for settlement time.

4.7 ELECTRONIC CHEQUES

The paper cheque is a very familiar payment instrument used by individuals, businesses and governments to pay for goods and services. For example, according to the US National Organization of Clearing Houses (NOCH), over 63.4 billion paper cheques were written in the United States in 1996.

The information on a paper cheque includes the names of the payer and the payee, the amount of the cheque, and the name of the paying bank. The Magnetic Ink Character Recognition (MICR) line at the bottom of the cheque includes the bank 'sort code' (identifying the payer's bank), the account number of the payer and a number that identifies the cheque itself. The MICR line permits cheques to be processed on high-speed equipment cheque processing equipment. Before such processing, the monetary amount of the cheque is also encoded in magnetic ink at the bottom of the cheque.

Worldwide, there are several ways of clearing cheques. For example, when cheques are deposited into the same bank on which they were drawn, these transactions can be settled in-house, and such cheques are referred to as 'on-us' cheques.

In the UK there is a single clearing system used by all banks, but in the US there are approximately 150 cheque clearing house associations, of which the three largest are California Bankers Clearing House (CBCH), the Chicago Clearing House Association (CCH), and the New York Clearing House Association (NYCHA).

In the last few years, some major US clearing house associations, the US banking industry, and the US Federal Reserve have been actively developing and pursuing a new technology that may shorten the amount of time it takes to clear and settle cheques, and thus improve the overall efficiency of the payment system. This new cheque technology is called Electronic Cheque Presentment (ECP). ECP is a process by which the MICR-line information is sent electronically to the paying bank. A number of large commercial banks participate in the Electronic Cheque Clearing House Organization (ECCHO), formed in 1990. ECCHO drafts rules and designs formats for electronic cheque processing among its members. Banks that are ECCHO participants can exchange electronic cheque data among themselves before the paper cheques are physically presented for payment.

ECP may include cheque truncation, and may be supported by cheque imaging technology. Cheque truncation is a process by which the paper cheques are retained at some point in the collection process, and only the cheque information is sent forward to the paying bank. ECCHO is developing a set of national rules for cheque truncation. Cheque imaging is a process by which a picture is taken of the front and back of the cheque, and the images are stored on electronic media for retrieval when needed.

Although the use of cheque truncation and imaging is steadily increasing, it is not clear how much cheque volume will be affected by these methods in the foreseeable future. The reluctance of some banks to invest in the technology, and consumer preference for their returned cheques, may restrain substantial growth in cheque truncation and imaging. One Federal Reserve official predicted that cheque truncation would not be widely used until consumers accepted the fact that their cheques would not be returned to them. Moreover, under current law, depository banks must physically present cheques to paying banks to obtain settlement for the cheques.

4.8 SET IN DETAIL

A more detailed description is given here of the SET architecture. Our reasoning here is that SET gives a good framework for understanding the issues of electronic payments and the main technology components that can meet them. Currently, it has not made much practical headway towards implementation in the UK, despite being backed by Visa and Mastercard. Some of the reasons for this are discussed later.

SET is a suite of technical specifications owned jointly by Visa and MasterCard. Technical input was provided by GTE, IBM, Microsoft,

Netscape, RSA, SAIC, Terisa and VeriSign. Version 1.0 of the specification was published in 1997. SET makes heavy use of security technologies, including certification and digital signatures. It strongly recommended that the sections of Chapter 5 relating to those topics are consulted prior to reading the following sections.

SET business requirements

SET identifies seven business requirements and proposes technical solutions for meeting them. These are:

1. *Confidentiality of information*
SET identifies the need to provide confidentiality of both payment information and the order information that accompanies it. This requirement is addressed through the encryption of messages. It is a significant point that the approach is to encrypt the messages themselves, rather than just securing the message transport mechanism. An encrypted transport such as secure sockets (see Section 5.3) is intended to provide privacy from eavesdroppers, but the recipient of messages can read them in their entirety. By encrypting the messages themselves, it is possible to pass information via several parties who act in different roles, and to ensure that each can only decode the appropriate messages that relate to their role. As an example, a merchant will need to decode the order information, but may not be able to do the same with the payment information that accompanies it.

2. *Integrity of all transmitted data*
All parties engaging in an electronic transaction must be confident that data has not been tampered with or modified in transit. SET uses digital signatures to sign the data before it is transferred. How digital signatures work is described in the chapter on security.

3. *Authenticate that a cardholder is a legitimate user of a branded payment card account.*
When someone makes a payment with a credit card it is important for the merchant to be able to associate the individual with a particular card account. In a 'cardholder present' transaction (e.g. in a shop), the merchant first links the cardholder to the card by requiring them to make a signature that matches that on the card, and then links the card to the account, probably by checking that the card looks genuine, and in many cases, by contacting the issuer to check that the account is valid.

Currently, many Internet transactions use the simple expedient of requiring a person to quote their card number, expiry date and some other piece of personal information, and this is considered adequate to associate the individual with the account, but this is not particularly secure and is open to abuse. For example, if I can obtain someone's credit card number and its expiry date (which I can accomplish from a quick visual inspection of their card) I can easily impersonate them in an electronic transaction.

SET addresses this issue by using a combination of Cardholder Certificates and electronic signatures. Cardholder Certificates are SET's eBusiness equivalent of the payment card. When these are issued they are digitally signed by the issuing bank, thus preventing them from being modified or forged (see section 5.2 on digital certificates). As noted above, in the case of a physical payments card, the 'secret' information it holds (which therefore authenticates its use in many practical circumstances) are the account number and expiry date. In the case of the Cardholder Certificate, the account information and a secret value (i.e. a PIN or password) are encoded into it using a one-way hashing algorithm. This ensures that these items can easily be checked against the contents of the Cardholder Certificate, but cannot be reverse-engineered from it.

4. Authenticate that a merchant can accept a particular branded payment card's transactions through its relationship with an acquiring bank
Cardholders need to confirm that a merchant has an account with an acquiring bank, thus allowing it to accept payment cards. The cardholders also need to be able to confirm merchants' identities so that, for example, if I am making a payment to Barclays Bank or British Gas, I am sure that the Internet site refers to *the* Barclays or BG companies, and not to some similarly-named site that is being operated by some less-than-scrupulous merchant. SET addresses this requirement through the use of digital signatures and Merchant Certificates. Merchant Certificates are the eBusiness equivalent of the Visa sticker that shops display in their windows, but have somewhat richer content. When Merchant Certificates are issued, they are signed by the acquirer, and hence they provide assurance that the merchant has a relationship with that particular acquirer.

5. Ensure the use of the best security practices and system design techniques to protect all legitimate parties in an electronic commerce transaction
As previously noted, SET provides a comprehensive architecture for payments. In terms of using best-practice, SET addressed this requirement by calling on the expertise of industry leaders such as GTE, IBM, Microsoft, Netscape, RSA, SAIC, Terisa and VeriSign.

6. Create a protocol that neither depends upon transport security mechanisms nor prevents their use
SET's intention is to build security into the applications and messages that pass between them, rather than relying on the existence of a totally secure transport infrastructure.

7. Facilitate and encourage interoperability among software and network providers
SET defines an open set of message specifications to encourage this. See SET Secure Electronic Transaction Specification Version 1.0 31 May 1997 for all the gory detail.

Example SET transaction

To give a flavour of the working of the SET architecture, we take the example of a cardholder buying some goods online and paying with a credit card. We have omitted some areas of complexity, such as the fact that all participants may be able to handle a variety of different payments cards from different vendors. We assume that payment will take place through a payments gateway operated by an acquiring bank. Furthermore, at the start of this we assume that some things are already in place:

- The cardholder already has an account with an Issuer and has received a Cardholder Certificate.
- The Merchant already has an account with an Acquirer and has received a Merchant Certificate.
- The Cardholder and Merchant each have a 'signing key' for signing messages. Each signing key is made of two parts: a public key and a private key (see chapter 5).
- The Payments Gateway has a certificate, a copy of which is held by the merchant. This certificate is effectively a signed copy of the Gateway's public key.

We join this transaction just as the cardholder has browsed the merchant's web site, selected items to purchase, has filled in some sort of online order form, and now needs to make a payment. The message flows and processing that takes place at the Cardholder's computer and Merchant's computer are summarised in the table below. The main message flows are as follows:

Initiate Request message

1. The Cardholder's computer sends an *Initiate Request* message to the Merchant's computer.

Initiate Response message

1. When the merchant's computer receives the *Initiate Request*, it first generates a unique Transaction Identifier that will be used to identify this thread of activity.
2. Next, it Generates an *Initiate Response* message (which includes the Transaction Identifier).
3. The merchant's computer digitally signs the *Initiate Response* message (by creating a digest and encrypting it with the merchant's private signing key).
4. The merchant's computer sends the signed *Initiate Response* message to the cardholder, together with the Merchant Certificate and Payments Gateway certificate.

Purchase Request message

1. When the cardholder computer receives the *Initiate Response* message, it first checks the integrity of the message by means of the Merchant's signature, i.e. the Cardholder's computer creates a digest of the message and compares this with the result of decrypting the signature, using the Merchant's public signing key. The two should be the same, proving that the message has not been tampered with in transit.
2. It verifies the Merchant and Gateway Certificates by traversing the trust chain to the root key. It thus gains assurance that the merchant and the gateway are indeed what they purport to be, and also that the Merchant's certificate has been issued by an authentic acquirer.
3. The Cardholder's computer next creates the Order Information (OI) and Payment Instructions (PI).
4. It generates a dual signature for the OI and the PI by computing the message digests of both, concatenating the two digests, computing the message digest of the result, and encrypting that using the cardholder's private signature key. This dual signature will be used to protect the integrity of the OI and PI individually and collectively.
5. The Cardholder's computer generates a random symmetric encryption key ('K') and uses this to encrypt the PI. This step is necessary to ensure the privacy of the payment information, particularly when this is handled by intermediaries such as the merchant.
6. The cardholder's account information is encrypted together with the value of 'K', using the public key of the payments gateway (taken from the Gateway Certificate).
7. The cardholder's computer generates a *Purchase Request* message which includes the OI, the encrypted PI, the message digest of the PI, and the dual signature.

Purchase Response message

1. When the merchant computer receives the *Purchase Request*, it verifies the cardholder signing certificate by traversing the trust chain to the root key.
2. Next it checks the integrity of the order, using the dual signature. This can be done by decrypting with the Cardholder's public key, and comparing the result with the concatenation of the OI digest (which the merchant can compute) and the PI digest (which was included in the Purchase Request message).
3. The Merchant's computer forwards the encrypted payment information to the Payment Gateway.
4. The order is processed by the Merchant in accordance with the contents of the OI.
5. The Merchant's computer generates and digitally signs a purchase response message. This is made up of a copy of the Merchant's signing certificate and an indication that the Cardholder's order has been received by the merchant.

Cardholder's Computer	Message Flows	Merchant's Computer
1. **Construct** *Initiate Request* message		
	Initiate Request	
		1. Assign unique transaction ID 2. Generate *Initiate Response* message (including Transaction ID) 3. Sign message with private signature key 4. Send message plus Merchant Certificate and Gateway Certificate
	Initiate Response +Merchant Certificate +Gateway Certificate	
1. Verify merchant's signature 2. Verify certificates by traversing trust chain 3. Create Order Information (OI) and Payment Instructions (PI) 4. Create a dual signature of the OI and PI (concatenate digests of each, create a digest of the result and encrypt with the Cardholder's private signing key) 5. Generate a random symmetric encryption key (K) and encrypt PI with it 6. Encrypt cardholders account information together with K, using Gateway's public key taken from its certificate 7. Create a *Purchase Request* message		
	Purchase Request +*OI* +*Encrypted PI*	

Cardholder's Computer	Message Flows	Merchant's Computer
		1. Verify Cardholder's signing certificate by traversing trust chain 2. Verify the dual signature 3. Forward encrypted PI to payments gateway 4. Process the Order Information 5. Create *Purchase Response* message and digitally sign it
	Purchase Response	
1. Verify Merchant Signature Certificate by traversing trust chain 2. Check integrity of message using digital signature		

6. When the Cardholder's computer receives the *Purchase Response* message from the Merchant, it first checks the integrity of the message by means of the Merchant's digital Signature. It then verifies the Merchant signing certificate by traversing the trust chain to the root key.

SET conclusion

Although there is a considerable degree of complexity in the above interactions, it can be seen that:

- The integrity of messages is verified at all stages through the use of digital signatures.
- Appropriate levels of privacy are maintained by the use of message encryption (e.g. the Merchant cannot see the detailed payment instructions that are sent to the Acquiring Bank).
- In spite of this separate treatment of order information and payment instructions, these pieces of information are integrated through the use of the 'dual key' that protects them as a combined entity whose integrity can be checked even by those (such as the merchant) who do not have access to all the contents.

- The identity of all parties is assured through the use of certificates (signed by trusted third parties).

At the time of writing, however, the SET standards have yet to make a major impact, and, for example, are not known to be in use anywhere in the UK. The reasons for this slow take-up are not entirely clear, but it is probable that a major factor is the need for the cardholder to own (and manage) a Cardholder Certificate and Signing Key. Whereas the management of plastic cards is well understood, both by issuers and cardholders, the same is not true of electronic certificates. For example, these electronic items might be left accessible on a computer that could be used by other people; copies could be made, intentionally or accidentally; and they are not conveniently portable in a secure manner (e.g. between home and work). It is likely that the main outstanding requirement for the adoption of SET is some form of physical container for electronic certificates, such as smartcards, and for facilities to be universally integrated into PCs for reading the data from them. This latter is not an insignificant issue, as it must be accomplished by trusted system components.

4.9 PAYMENTS, MICROPAYMENTS AND NANOPAYMENTS

When dealing with online transactions, we have to be able to deal with various scales of payments: for instance, items such as books are regularly purchased electronically (e.g. through Amazon.com), and it is common for a single customer to buy a single book and pay with a single credit card transaction (probably in the $10+ range). However, the cost of processing credit card transactions is quite high, and may not be economical for the purchase of small-value items such as consumer reports or individual music tracks, which would typically be in the $1 to $10 range. Technologies exist for handling these small-value payments (known as 'micropayments') and lumping them together into single credit card transactions. The range below $1 is often referred to as the area of 'nanopayments'. Such payments may represent, for example, the price of viewing an individual Internet page of information—containing a share price, for instance.

The aggregation of payments may, in principle, take place either at the server side or at the client side. Current systems, though, are generally server-based because of the difficulty of securing client-side payment information. For example, simple client-side payment aggregation systems that store details of transactions that have been made can be defeated simply by deleting the file in question. Hence, the usual approach is to maintain a user account at the server and to record every transaction made by that individual user. However, client-side systems, whilst more complex to implement, offer some advantages: they are better suited to applications that will be used by extremely large numbers of users (because all the account information for all the customers does not have to be held in a central location), they allow options such as 'electronic wallets' which are portable by the customer, and perhaps most importantly, they lend themselves to payments systems that are general-purpose, rather than being associated with a particular business application. For example, the same client-side payments aggregation engine may be recording costs incurred from the use of many different internet services. Some architectures for client-side payments systems (e.g. that developed by Intertrust) do exist but have yet to make a major impact on the buying public.

As well as the choice between client and server-based aggregation systems, the other choice is between pre-payment and post-payment. In other words, if you are offering customers many small-scale 'products' for which they will aggregate micropayments, do you require them to pay for a minimum number of transactions before they begin (and then keep a record of how much 'credit' they use up), or do you record the details of every transaction and wait until the customer reaches a certain threshold before charging them.

An example of a prepayment system can be found at the web site of the Scottish Register Office (www.origins.net). This online system allows members of the public to search for birth, marriage and death certificates and then to order them. Before being permitted to make a search, customers must register and make a payment from their credit card. At the time of writing the minimum charge for making searches is £6. On making this online payment, customers are credited with 30 'screens-worth' of searches. They can then begin searching for the event they require: for example, all the people called Hamish Robert Brown whose births were registered in December 1851. Each such search returns results which occupy one or more screens, and every time a screen is viewed, the system decrements the number of screen credits. If, eventually, the full 30 screens have been viewed and the counter is decremented to zero, the customer is prompted to re-enter their credit card details and make a further payment.

An example of a post-payment system is BT Array. This offers a number of information sources for sale, including reports from the consumer magazine *Which?* When customers register on this system, they must provide their credit card details. Then, they may purchase items and their account on the system will be debited (although no actual financial transaction will take place yet). When their account total (i.e. the amount they owe) reaches a certain threshold (currently £30), a charge is made against their credit card to clear their account. In practice, such a credit card payment is also made if a certain time limit is reached.

	Pre-pay	Post-pay
Summary description	Consumer is charged in advance of purchases. Account is debited every time a purchase is made	Cost of purchases is added up until a threshold (and/ or time limit) is reached, at which point the payment is made
Advantages	Less risk to merchant (credit card can be charged before goods are purchased)	
Disadvantages	Potential customers may be deterred by up-front costs. People who purchase credits but then make few purchases may be dissatisfied with the service	Problems with credit card may arise when payment is due (e.g. expired cards) Credit card details need to be stored (resulting in additional security requirements for merchant) In some jurisdictions this may be construed as providing credit and may therefore bring the merchant within the terms of consumer credit legislation

The benefits and drawbacks of pre-pay versus post-pay are summarised in the table below.

Not only are different technologies required to aggregate these various categories of small-scale payments, but also the eBusiness will need to put in place different strategies for such things as handling account queries and dealing with bad debts (in the case of nanopayments, the debts would probably be written off and the offenders blacklisted, as the cost of recovering the debt is too high).

4.10 ELECTRONIC FUNDS TRANSFER

At the start of the chapter, we said that we would be concentrating on the links between cardholder and merchant and between the merchant and their acquiring bank. Up to this point we have done so, and have neglected the interactions that take place between the financial institutions themselves, such as those that take place between the cardholder's issuing bank and the merchant's acquiring bank. These 'behind-the-scenes' interactions are the domain of Electronic Funds Transfer (EFT). Three of the major systems in current use are SWIFT, which operates as a global system, Fedwire and CHIPS, which are the two major US-based systems.

The Federal Reserve's Fedwire funds transfer service, primarily used for U.S. domestic payments, is a real-time, gross settlement system in which the Federal Reserve guarantees payment to the receiver of the funds. CHIPS, a private-sector multilateral netting organisation, is used mainly to settle the U.S. dollar side of foreign exchange transactions. In 1996, Fedwire funds transfer service's average daily transaction value was \$989 billion, and the average amount per transaction was \$3.0 million. CHIPS' average total daily transaction value was \$1.3 trillion, and the average value per transaction was \$6.2 million.

SWIFT, Fedwire and CHIPS are described in more detail below.

SWIFT

The Society for Worldwide Interbank Financial Telecommunication (SWIFT), incorporated in Belgium, is a co-operative owned by over 2800 banks from around the world, including over 150 from the United States. It was founded in the early 1970s, and operates a network that processes and transmits financial messages among members and other users in 137 countries. At the time of writing, it supports over 6000 financial institutions in 178 countries. In 1998, SWIFT's global network carried over 900 million messages whose average daily value was above 2 trillion US dollars.

Today, in addition to its 2,800 member banks, SWIFT users include both sub-members and participants such as brokers, investment managers, securities deposit and clearing organisations, and stock exchanges.

SWIFT messages convey information or instructions between financial institutions: messages are formatted and contain information about the originator, purpose, destination, terms and recipient. The largest use of SWIFT is for payment messages, through which one institution transmits instructions to another to make payments. Other messages are used to confirm the details of a contract entered into between two users, such as a foreign exchange trade or an interbank deposit placement.

For securities, SWIFT messages can transmit orders to buy or sell, or convey instructions concerning delivery and settlement.

SWIFT infrastructure

The SWIFT network is based on the X.25 packet-switching protocol running over a collection of leased and dial-up lines. Customers (e.g. banks and financial institutions) attach to this network via local 'access points' (i.e. points of presence) to which they are connected by leased lines or, in certain countries, through Public Switched Public Data Networks such as Transpac, Accunet and Sprintnet. At their premises, customers must operate an approved SWIFT interface between their own systems and the network. Such interfaces are available from SWIFT themselves, and from third parties.

The message processing system consists of a network of servers which route messages and maintain an audit trail as the messages flow through the network. The computing systems are duplicated in two operating centres (one in the US and the other in the Netherlands) to provide resilience.

The operating centres consist of three types of processing components:

- *Regional processors* are the 'outer layer' of processing. They control the traffic flow to and from the access points, and they monitor message entry and exit from the system, and validate message syntax and integrity. They forward messages into the 'inner layer' of so-called slice processors.
- *Slice processors* make up the core store-and-forward infrastructure. In addition, they create copies of messages (for audit purposes) and generate confirmations of message reception to the sender.
- *System control processors* manage the entire system: they monitor the performance, administer the network, and support maintenance tasks.

The SWIFT network has been designed to protect against a range of identified security threats such as malicious or fraudulent use of the network, attacks against the physical resources or environmental disaster. Some examples of the security features are:

- *Private network*: SWIFT is based on a private network (rather than, for example, the public Internet), and is therefore less prone to attack from casual Internet hackers.
- *Message validation*: messages are checked for compliance with the syntactical rules and message formats: the presence of mandatory fields, valid message tag sequences, order, etc.
- *Message sequence checking*: all SWIFT messages are assigned unique sequence numbers when entering and exiting the system. These numbers are verified during sending and reception and those that do not follow the expected sequence are rejected, and their associated terminal session is also aborted.
- *Message integrity*: when a message is input to the SWIFT network, a 'Message Authentication Code' is created, based on the message content. This is similar to the 'message digests' discussed in Section 5.2, and hence provides assurance that messages are secure from accidental (or deliberate) alterations during the message transfer process.

- *Strong log-in authentication*: this is based on smart cards and the exchange of electronic authentication keys.

- *Data encryption*: data is encrypted through the network from access point to access point, and optionally, to a user's interface device. Hence messages are protected from viewing by SWIFT's employees whilst in transit or in storage in the system.

SWIFT quality of service

SWIFT is a near real-time service. Although it is based on a store-and-forward architecture, messages and file transactions are processed immediately and, if the sender and receiver are both connected online, a message transfer typically takes less than 20 seconds, even for international message transfers.

SWIFT provides a financial guarantee for the delivery of all messages sent over its network. SWIFT has a Responsibilities and Liability Policy, under which it assumes financial liability for direct interest losses resulting from authenticated payments or transfers that it fails to deliver.

There is a full audit trail, and the system is available 24 hours a day, 7 days a week.

SWIFT applications and services

The core message store-and-forward service is branded as *FIN*. There are a number of enhancements to this basic service, such as *FIN Copy* (which forwards message copies to third parties, to support certain financial transactions such as clearing, netting and settlement of payments) and *FINInform* (which enables messages to be copied to a third party within the same financial institution).

A *Bulk File Transfer* service provides store-and-forward facilities for unstructured files containing arbitrary amounts of data. This service is based on X.400 messaging standards.

Other services are specific to the financial industry, such as *Matching* (automatically validating and reconciling contracts) and *Netting*.

The financial industry has the concept of Straight Through Processing (STP), i.e. the ability to automate end-to-end processes without manual intervention. The structured messages employed in SWIFT lend themselves to automated parsing and processing. SWIFT assists this by offering an STP Traffic Analysis Service to distinguish 'clean' messages from those that are likely to cause problems in an automated terminal.

Message formats

The main current international standard in this area is the ISO securities message standard, ISO 15022 'Securities—Scheme for Messages (Data Field Dictionary)'. This replaces two previous International Standards for electronic messages exchanged between securities industry players: ISO 7775—'Scheme for Message Types' and ISO 11521—'Scheme for Interdepository Message Types'.

The ISO 15022 standard consists of a set of syntax and message design rules, a dictionary of data fields and a catalogue for present and future messages. However, the contents of the Data Field Dictionary and Catalogue of Messages have been kept outside the standard, but are controlled by a Registration Authority. ISO has appointed SWIFT as the Registration Authority for ISO 15022.

SWIFT is bringing its message set into compliance with ISO 15022, and is registering the detailed data fields with the Registration Authority.

The other existing standard for messages is that controlled by EDIFACT (Electronic Data Interchange for Administration, Commerce and Transport). The SWIFT service will transport the EDIFACT standard payments messages.

SWIFT messages are divided into ten main categories:

0xx General Information
1xx Customer Transfers
2xx Financial Institutions Transfers
3xx Financial Trading
4xx Collections and Cash Letters
5xx Financial Trading (Securities)
6xx Precious Metals Trading and Syndications
7xx Documentary Credits and Guarantees
8xx Travellers' Cheques
9xx Cash Management and Customer Status

Fedwire

The Fedwire funds transfer system is one of the two primary large-dollar electronic payments systems in the United States. In 1996, Fedwire's average total daily transaction value for the electronic transfer of funds was about $989 billion, and the average value per transaction was $3.0 million. The Fedwire funds transfer service allows depository institutions to transfer funds on their own behalf or on behalf of their customers; most Fedwire payments are related to domestic transactions. The Department of the Treasury and other Federal agencies also use Fedwire to disburse and collect funds.

Fedwire is a Real-Time Gross Settlement (RTGS) system. It consists of two components: a high-speed, US-wide communications network (FEDNET) that electronically links all Federal Reserve Banks and branches with depository institutions; and the set of computers that process and record individual funds transfers as they occur.

Fedwire's computer centre (the East Rutherford Operations Center—EROC) operates a main computer system, with a hot-standby backup system that can take over almost instantaneously in the event of catastrophic failure of the main system. In addition, there is a secondary back-up centre (Federal Reserve Bank of Richmond) which can take over with 60 to 90 minutes of a total failure of the entire primary site, and a tertiary centre that can take over if both the others fail. Disaster protection is a major consideration in this system!

A variety of data security measures protect the integrity, confidentiality and continuity of the system. These include access, authentication and verification controls; data encryption; procedural controls over such processes as application changes, data entry and database updates; physical security; and personnel standards. These controls are designed to prevent tampering with, destroying or disclosing Fedwire data, either by Federal Reserve employees or outside hackers. For example, Fedwire messages between depository institutions and the Federal Reserve are encrypted and authenticated to prevent interception and alteration. Access controls, such as unique user identification codes and passwords, are also a primary means for preventing unauthorised transfers. For example, employees at a depository institution must use a valid and unique user identification code and password to enter and send a Fedwire message, and that message must come from a connection associated with that employee's institution.

A critical feature of Fedwire is that it offers immediate 'finality' (i.e. final and irrevocable credit) to the recipient. The Federal Reserve 'guarantees' the payment to the depository institution receiving the Fedwire transaction, and assumes any credit risk if there are insufficient funds in the Federal Reserve account of the bank sending the payment.

Clearing House Interbank Payment System (CHIPS)

The Clearing House Interbank Payments System (CHIPS) is the other large-value electronic payments system in the United States. It is privately owned and operated by the New York Clearing House Association (NYCHA). It began operations in 1970 as an electronic replacement for paper cheques in international dollar payments. Although Fedwire payments are principally related to domestic transactions, US dollar payments related to 'foreign transactions' (such as the dollar leg of foreign exchange and Eurodollar placements) flow primarily through CHIPS.

Although CHIPS transfers are irrevocable, they are final only after the completion of end-of-day settlement. CHIPS nets its transactions on a multilateral basis. Thus, if a bank receiving a CHIPS transfer makes funds available to its customers before settlement is complete at the end of the day, it is exposed to some risk of loss if CHIPS does not settle. However, in its 27 years of operation, CHIPS has *never* failed to settle. (This contrasts with Fedwire, which offers immediate finality of settlement—the Federal Reserve 'guarantees' the payment and assumes any credit risk.)

To illustrate how CHIPS transactions work, we take the example of a European merchant who wishes to pay $2 million to a US supplier for a shipment of consumer goods. The merchant instructs his bank to debit the Euro equivalent of $2 million from his account and to arrange payment of the dollar amount to the US supplier's account in the USA. The transaction between the two banks is not settled until CHIPS settles at the end of the day. Hence, technically, the US bank would be exposing itself to risk if it made the $2 million available to its customer during the day of the transaction.

In a typical day, CHIPS participants may continuously exchange payments amongst themselves, and CHIPS continuously recalculates each participant's single net position *vis-à-vis* all other participants combined. This is called 'multilateral netting'.

Settling participants take part in the actual settlement of CHIPS by sending or receiving the Fedwire payments used to effect settlement. Participants that are not settling participants must designate a settling participant to settle for them, and that settling participant must agree to the designation.

After all the settling participants that are in a net debit position have paid in funds and participants that are in a net credit position receive a Fedwire funds transfer from NYCHA, the CHIPS settlement account at the Federal Reserve Bank of New York (FRBNY) reaches a zero balance. It is at this point that the transaction transmitted over CHIPS between participants is settled, and settlement is final.

CHIPS maintains a warm-standby backup centre which can resume payment processing within five minutes of a failure of the main processing centre. If the CHIPS database suffers damage, CHIPS has a computerised method for rebuilding its database. Each participant can automatically retransmit a payment previously sent if CHIPS indicates the loss of payments through the transmission of the recovery report. According to an NYCHA document, CHIPS quarterly tests its contingency plans against a variety of simulated events in mandatory exercises involving all participants.

4.11 SECURITY AND PROTECTION FOR NEW FINANCIAL PRODUCTS AND SERVICES

Electronic technologies now in place or under development, such as the use of the Internet for financial transactions, electronic cash, and stored-value cards, hold great promise for increasing consumer choice in payment methods. However, such technologies also provide additional means for abuse and illegal activity. Law enforcement officials have expressed concerns about the possibility of individuals using these new products and services for illegal purposes, such as money laundering. Some fraudulent schemes involving securities transactions over the Internet have already been uncovered. Because of the newness of these products and services, approaches to making such products and services more secure and less vulnerable to illegal use are still under development.

Through the Internet, customers now have access to credit card payment systems, electronic banking, and other financial services in a way that was never before possible. Such services bring to the consumer an array of new, convenient methods for doing financial transactions. However, because of the nature of the Internet—the fact that it is basically an unsecured means of transmitting information—customers, merchants and other service providers have increasing concerns about the safety and security of their transactions. For example, if an intruder could successfully attack a credit card association, customers could lose access to their accounts, or in the worst case, the credit card payment system would grind to a halt.

Firewalls and other methods of filtering information coming from the Internet to computers can be used, in part, to increase the security of Internet transactions. A firewall is a method that attempts to block intruders by limiting the information that can pass to the merchant's or the financial institution's internal network. Use of encryption is another means by which the security of Internet transactions could be increased. However, neither firewalls nor encryption provides a guarantee of safety for Internet transactions. To the extent that financial transactions over the Internet remain vulnerable to intrusion or capture by unauthorised parties, consumers may be reluctant to dramatically increase their use of the Internet for their financial business, and any major successful attacks would probably affect the public confidence in electronic commerce in general.

In addition to the issue of the basic security the Internet also provides a new means for criminal elements to perpetuate fraudulent schemes against consumers. Such schemes pose risks to consumers, because the Internet provides relatively easy and cost-effective access to millions of individuals. Pennsylvania securities regulators have described several illegal schemes conducted over the Internet, including sales of non-existent bonds. Law enforcement agencies are stepping up their efforts to identify and stop such schemes, but it remains to be seen whether these efforts can keep pace with the rapid growth of the use of electronic commerce and the Internet for such illicit purposes.

In addition, new technologies, such as the use of the Internet for financial transactions or stored-value cards, are likely to provide additional avenues for money laundering. Law enforcement officials are especially concerned with systems that allow person-to-person transfers, which would include stored-value cards and Internet transfers. Stored-value cards may enable individuals to move illegal money from a bank account onto a stored-value card, where it will be untraceable when used. However, because stored-value cards generally are designed for small-value purchases and many have low limits, such as $500, for the amounts that can be stored on them, they may not be very efficient vehicles for laundering large amounts of cash.

4.12 HOW NOT TO BECOME A BANK OR CREDIT COMPANY

Whilst you may want to conduct business electronically, you probably don't want to become a bank! (and you probably don't want to find yourself subject to banking law or consumer credit law). Although legal advice is well beyond the scope of this book (and, in any case, is country-dependent), there are some definite recommendations of things to avoid, and some of the things you can and cannot do with other people's money. Also, electronic payments are an area that is being covered by a number of European directives.

Some of the main areas of legislation that apply in the UK are:

- The 1987 Banking Act. Amongst other things, this only allows authorised institutions (such as banks) to provide facilities for depositing money. Pre-payment for goods does not count as taking deposits.
- The Consumer Credit Act, which applies in many cases where a payment system gives credit to consumers.
- The EU data protection directive.
- The (proposed) EU electronic signatures directive.
- The EU 'Proposal for a European Parliament and Council directive on certain legal aspects of electronic commerce in the international market' (published in November 1998).

4.13 ELECTRONIC BANKING

The ever increasing use of the Internet and the ubiquity of the PC set the scene for banks and other institutions to offer electronic banking services.

In addition to delivery through the PC, such services may also be delivered by means of automated teller machines, mobile phones, TV set-top boxes and specialised terminals.

Some of the main services offered by electronic banking include:

- Bill payment
- Funds transfer between personal accounts
- Balance enquiries
- Share dealing
- Loan applications
- Foreign currency applications.

Banks and other financial institutions, as well as new entrants into the market such as supermarkets, are now competing to offer electronic banking services.

In the UK, the 'traditional' high-street banks have perhaps been reluctant to embrace the possibilities of electronic banking because they perceive this development as a threat to their conventional business. In the pre-electronic-banking era, it was very difficult for new entrants to make inroads into the UK banking market without making a huge up-front investment in high street premises in major towns. Electronic banking changes all that, and makes it feasible for new banking institutions to enter the market with only a virtual presence in the market.

4.14 STANDARDISATION

The work of standards organisations tackling the issues of trust and security for the payments industry have consolidated their activities in four main areas:

- *Payment protocols*: define how two or more of the e-commerce roles in a transaction exchange messages so that a payment can occur. Cardholder, merchant and acquirer authentication are handled by one or more of these protocols, as well as confidentiality and integrity of payment data. The key emerging standards to track/follow/adopt in this field are SET, EMV, Mondex, and Visa Cash. SSL is currently widely used for payments on the Internet, particularly because it is built into every browser, but it lacks the more comprehensive security framework of the other standards.
- *Trading/shopping protocols*: broader in scope than payment protocols (but

able to include them too), they address events that are integral to the end to end purchase transaction, including negotiation of terms, invoicing, receipting, taxation, delivery, etc. The most important of these are the Internet Open Trading Protocol (IOTP), now adopted by the Internet Engineering Task Force (IETF), and the Open Buying on the Internet (OBI) protocol, aimed principally at the business-to-business sector and described as an 'EDI meets the Net' effort. It is likely that both the UK and German governments will adopt the IOTP for purchase tax transactions on the Internet.

- *Reader standards and operating systems for smart cards*: the smart card industry has suffered from the proliferation of proprietary solutions, usually promoted by the once powerful card manufacturers. However, with the interest of the software and IT industry in smart cards both Microsoft (PC/SC) and SUN/IBM (Opencard) are promoting 'open' solutions for reader interfaces. On the operating system front, Microsoft announced the 'Smart Card for Windows' operating system in October 1998, a move expected to provide strong competition to the current industry leader MULTOS (owned essentially by the Mastercard Mondex shareholders).
- *IP based security*: 'IP Sec' standard sponsored by the IETF, the output of which is covered in the next chapter.

There are several industry consortia producing specifications both for generic trading and for specific payment protocols. These consortia are a vital enabler for eBusiness as the current proliferation of proprietary solutions is often makes retailers loathe to invest in something that may have limited shelf life. In addition to the SET protocol detailed above, the other key consortia at present appear to be:

- *The Internet Engineering Task Force*: the IETF have now taken over the OTP protocol from the original OTP consortium, in which Mondex were a key player. As Mondex is a true payment only protocol, there need to be higher level trading protocols that provide the environment in which the payment can take place. These higher level protocols cover issues such as negotiation of terms, invoicing, receipting and delivery, etc. As some of the main providers of electronic payment systems (Cybercash, Mondex, Globe ID) were involved in the original OTP specifications, it is likely that the non-payment elements of their systems will migrate to a common set of protocols.
- *World Wide Web Consortium*: the W3C has been developing a MicroPayment Transfer Protocol (MPTP) since 1995. There has also been recent activity in the W3C micropayment group to define a payment API. This work is predicated on the existence of a wallet on the users client,

but with the expected use of XML this could have an impact in that it may define a standard payments tag that can be embedded in the HTML pages that a retailer publishes. This tag would then be interpreted by the client to initiate a payment using an appropriate payment mechanism.

- *Open Financial Exchange*: the OFX standard grew out of a joint initiative by CheckFree, Intuit and Microsoft in 1996. OFX bill presentation is a part of the standard, which allows large organisations to develop a system that can interface with partners and bill consolidators. CheckFree has set up an E-Bill partner program to encourage wider adoption of Internet billing. It currently offers e-billing and payment services to over two million customers in the US, and electronic links to 1000 merchants. AT&T joined with CheckFree in mid-1998 to offer residential customers the ability to view and pay their bills over the Internet using CheckFree's electronic systems.

4.15 SUMMARY

The technologies for electronic payments have advanced to the stage of being mature and well understood. Many companies are now offering solutions, predominantly based on card schemes, into the marketplace. At the same time that technology is enabling online payments, user suspicion is subsiding. The general awareness of the security available on the Internet means that reticence to key-in credit card details over the Internet is waning.

This chapter has set out the main options for paying for goods and services over the Internet, and has examined, in some detail, the technical basis for each one. The key components of online payment systems, such as SET, as examined in some detail and the operation of established instances, such as SWIFT, are described.

Progress in integration between catalogue suppliers and payment suppliers mean that it is becoming much easier for retailers to set up web sites that can handle online payments. Future schemes offering different payment mechanisms are likely to be released into the marketplace as new and emerging business models demand payments schemes not based on using cards.

FURTHER INFORMATION

SET at http://www.setco.org
OTP at http://www.otp.org

5

Trust and Security

We should distrust any enterprise that requires new clothes

Henry Thoreau

Walking down the high street your decision to go into a shop would be based on its location, size, type of premises, how long it has been there, and so on. If you are happy with what you see, you may well hand over cash in return for goods which you carry away. The risks are very small, and you can have a lot of confidence in what you are doing. Even if things go wrong, you know where to go back to and have every expectation of being fairly treated.

When trading over the Internet, things are not so simple. The delights of on-line shopping can quickly dissolve. For example, how does the consumer know (before handing over credit card details) that the company is reputable and is what it purports to be? What assurances are there that the company will be there tomorrow? Conversely, how does the trader know that the consumer is not using stolen credit card details?

Furthermore, how do both parties ensure that their transaction takes place privately without someone else snooping on it or, even worse, tinkering with the transaction details whilst they are in transit across the network? If things do not go smoothly, what sort of evidence is there to support both parties in their quest for a fair settlement?

Trust and security are two of the most important issues in any online transaction. They are cited as the most significant barriers to eBusiness. Not surprising, really, when you consider that two parties who never meet and who have no idea where each other is located are trying to strike a deal; and increasingly, one with significant value.

There are several aspects to trust and security. At the most basic level is the ability to ensure that communication over a network is protected. This is where encryption plays an important role in online transactions—no amount of subsequent work can make up for an insecure connection. Once the dialogue between customer and supplier is secured, other issues come to the fore. The identity of both parties needs to be assured before they can trade in good faith. Finally, each participant in a transaction will want assurance that the other party is trustworthy.

Another concept that is frequently encountered is that of 'non repudiation'. For example, under English law, if you can prove that you sent someone a letter through the postal system, that is taken as proof of its delivery (or at least, the onus is on the recipient to prove that it was not delivered). In eBusiness, there is no such legal precedent, so we need mechanisms that enable us to prove when, and to whom, a particular item was delivered.

So, there are many aspects to trust and security. These are all covered in this chapter, starting with the technical mechanics of securing a communications link, moving on to the issue online parties proving that they are who they say they are, and finishing with the softer topic of establishing trust.

5.1 PRIVACY AND ENCRYPTION

Secret codes have long been used to ensure the privacy of information that must be sent by untrusted carriers, such as a public telecommunication network. Some of the earliest codes can be traced back to the Ancient Egyptians. In 1660, Samuel Pepys was certainly devising such codes for use by English aristocrats communicating with the exiled King Charles. By the 1940s, highly sophisticated codes were being generated by the German 'enigma' machine to protect wartime secrets.

In eBusiness, the principle is much the same, although the coding (or cryptography) must be a lot stronger because the potential eavesdroppers are today very sophisticated, and must be assumed to have access to powerful computers and software. This section describes the main techniques used in current products such as web servers and browsers. Also of concern here are the constraints imposed by the governments of various countries: for example, the US will only permit the export of 'weak' cryptographic products; in France, cryptography has recently been all but illegal, and so on. First, we consider what a good security scheme should cater for.

Here be monsters

Network security is all about safeguarding the operation and preserving integrity in the face of accidental damage or deliberate attack. There are many aspects to security, from privacy (the ability to keep secrets) and integrity through to the '3As' of Authentication (knowing who people are), Authority (allowing them to do only what they should) and Audit (the 'forensic' trace, to see what happened, who did it and when). The main threats that a good security scheme should be capable of repelling are:

- Direct Attack—attacker aims to log on to an application and use it as though they were a legitimate user, but for covert purposes. The attack might involve stealing or guessing passwords, using operating system or application 'backdoors' (unsuspecting users may download and run software containing a Trojan Horse which introduces such a backdoor—for example 'Back Orifice' is used to attack PCs in this way) or subverting user authentication procedures.
- Denial of Service—prevents a network or application (e.g. a web server) from operating correctly. There are plenty of options to assist hackers in their attacks, and they could go for bandwidth (using devices such as the ping of death), server connection (inhibiting access to TCP/IP sockets using TCP SYN), etc.
- Loss of privacy—data is tapped in transit, and this data could be used to damage the company's reputation or for criminal purposes. LAN sniffers and WAN datascopes are readily available, and can be used for such illicit purposes.
- Data Modification—data in transit could be modified (e.g. a purchase of £1000 could be made, but the attacker could modify the data to show that only £10 had been spent).
- Masquerade—the attacker simply pretends to be the legitimate host. This could be a web site with a similar URL, designed to defame a company or gather money under a false pretence.
- Information gathering—often the prelude to one of the above attacks. Sophisticated scanning tools can be used to systematically search a host for security vulnerabilities (e.g. SATAN). Indeed, many rather effective hacking tools can be freely downloaded from the Internet!

In short, there are plenty of different ways in which a transaction over a public network can be compromised. Fortunately, there are also many ways to counter the threat, and many suppliers of security advice and equipment. The main guard is an effective mechanism for coding the information sent over the Internet.

Encryption

The primary role of encryption is to ensure the privacy of data when in transit across a public network such as the Internet. So, one of the most effective ways to protect a network is to encrypt all the data flowing over it, leaving no intelligible data available to the hacker.

Most schoolchildren have at least a passing acquaintance with encryption— they will have used some form of code to disguise messages passed around their friends. A typical example of this is shown in the top half of Figure 5.1. The transposition of letters (very simple in this case) is a very basic secret algorithm. Exactly the same idea can be applied to digital data. For example, the Clipper chip designed by the US Military contained an encryption algorithm known as Skipjack. Any two 'clipper-enabled' devices could exchange data, and with considerably more security than with our simple example!

The only drawback is that all data is visible if the secret algorithm is discovered. If the algorithm is burnt into silicon, as is the case with Clipper, it is difficult to restore security.

A more flexible option is to use an algorithm that can be specialised with a secret key, as illustrated in the lower half of Figure 5.1. The key, which is just a string of digits, contains the rules that drive the encryption algorithm. Hence, the issue now is to protect the key, rather than the algorithm, so the way in which the key (or keys) are managed needs to be considered carefully.

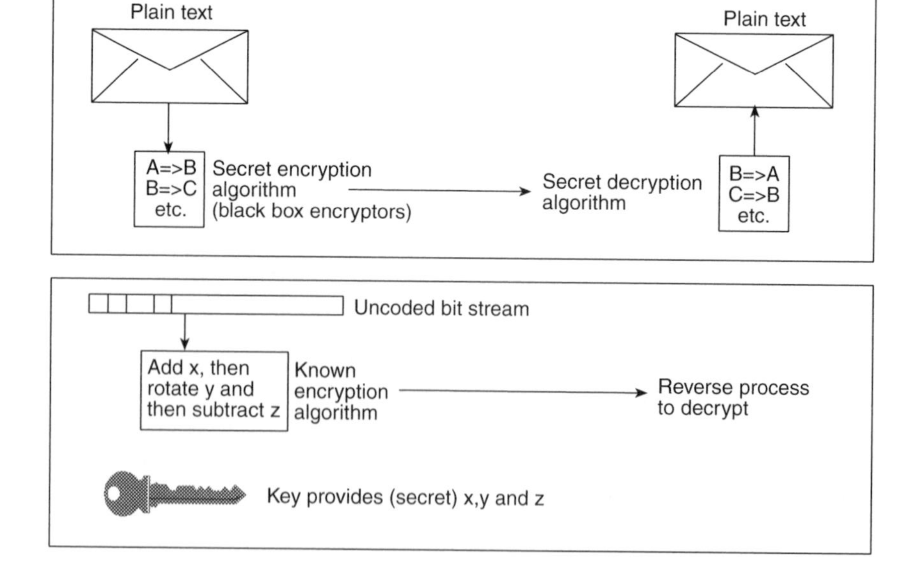

Figure 5.1

There are two basic approaches to key management. The first is known as 'private key cryptography', and is a system where the security of the encryption depends upon a shared secret that only the two communicating parties know. This shared secret is the private key that is used to encrypt data at the sending end and to decrypt it at the receiving end. The International Data Encryption Algorithm (IDEA) and Data Encryption Standard (DES) are examples of private key systems. Private key cryptography is also known as 'symmetric cryptography', owing to the symmetry of the encryption and decryption processes.

The long-standing standard for cryptography is DES. This is a symmetric private key system that uses a 56 bit key. This key is extended to 64 bits by the addition of 8 parity bits, and then this is used to encrypt data, which itself is broken up into 64 bit blocks. The algorithm takes these 64-bit blocks of data and combines them with the key through a complex sequence of bit-level operations such as exclusive-OR operations combined with bit shifting.

Originally, this algorithm was designed for implementing in custom hardware that could automate the bit shifting and comparisons very efficiently. However, nowadays, the same can easily be accomplished in software. The 56 bit DES algorithm remains the dominant approach to cryptography in some industries, particularly in the finance sector. However, in practice, computer technology now exists to break 56 bit DES in remarkably little time, and there is increasing demand to use stronger encryption, a point illustrated later in this section.

The alternative approach is known as 'public key cryptography', and is a system where each user has a pair of keys—one private and one public. A message encrypted using the public key can only be decrypted using the private key, and *vice versa*, so you can receive messages from anyone who knows your public key (which you decrypt with your private key) and can happily send an encrypted message to anyone whose public key you know. Public key cryptography is also known as 'asymmetric cryptography', because of the different keys involved in the encryption and decryption processes.

There are two ways in which a public key systems can be deployed—it can provide both

- *confidentiality*—the public key encrypts a document such that it can only be read by the owner of the corresponding private key. This approach is often used as a secure way of exchanging symmetric keys, and

- *digital signatures*—documents signed using the signer's private key may be validated using the corresponding public key. This provides a basis for authentication, integrity and non-repudiation and is explained in more detail below.

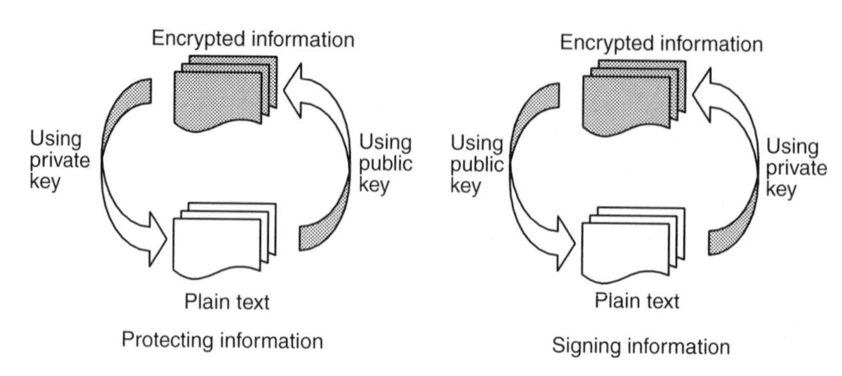

Figure 5.2 The use of public and private keys

The two uses of the public key approach are illustrated in Figure 5.2—the order of key application determining whether data is being encrypted or a digital signature decrypted.

Perhaps the best known of the public key systems is RSA, which is based on the use of very large prime numbers to generate keys that are resistant to malicious attack (i.e. being cracked). RSA exploits the fact that, faced with the product of two large prime numbers, it is mathematically very difficult to work out what those two prime numbers are.

In terms of usage, the private key systems, which use symmetric algorithms, are the more popular. Because these use the same key for encryption and decryption, they operate on blocks of data and subject them to a pattern of bit permutations and logic operations controlled by the key value. As such, they are able to operate at high speed and can be implemented using silicon hardware. Public key systems tend to be quite slow, so are used only for specialist applications (e.g. protecting the transmission of private keys and digital signing).

For all it's worth, there are some issues with encryption. The main one is actually obtaining the necessary hardware or software that removes the (sometimes) significant processing burden in encoding and decoding the data. Special purpose chips (such as Clipper) have been developed to do this but, because encryptors can have military use, their sale and export is rigorously controlled. Perhaps the best known example of such a restriction was the US Government's Data Encryption Standard (DES), which was classed as 'munitions' and thus subject to export restrictions.

A second issue with encryption is the speed with which the technology advances—codes considered secure but a few years ago can be readily broken within minutes using the latest processors. Because the effectiveness of both public and private key cryptography depends upon keeping keys secret, they must be immune from attack. The key's level of immunity is governed by the length (i.e. number of bits) of the key that is used to encrypt the data. However good the encryption algorithm, it can easily be defeated if the key can be guessed; and the shorter the key, the easier it becomes for someone to guess it. In practice, 'guessing' the key may entail trying every possible key until the right one is found. This is much like trying to open a combination lock by working through every possible combination. Such a 'brute force' approach is not feasible for an individual to perform but, given a powerful computer that may, perhaps, make millions of attempts per second, it can become a straightforward problem to crack, as we now illustrate.

We've stated that the difficulty of cracking an encryption algorithm increases with the number of bits used in the key. The general rule is that the time to crack is 2 to the power of the number of key bits minus the number of key attempts, all divided by 2. Or $2^{\text{key length-attack rate}}/2$. We can illustrate just what this means with a practical example. A US-based finance organisation recently built a hardware DES cracker, enticingly called 'Deep Crack', for the modest investment of $210,000. Deep Crack was capable of testing around 92,160,000,000 keys/sec, which is just under 2^{47} keys/sec in our preferred notation. This is an impressive performance, and one that well outstrips the typing capability of the average superhighwayman.

The algorithm under attack (DES, on this occasion) has the feature of a rather simple key schedule which considerably simplifies the design of a cracking machine. Hence, you would expect a rapid result—and, indeed, you get one! Applying our known 'time to break the code' formula, 56 bit DES fell victim to Deep Crack in $2^{56-47}/2$ seconds, a little under five minutes; and this assumes that we didn't get lucky and find the key on the first attempt.

Should we be worried about this? Are the sceptics right to say that security over the Internet simply isn't up to the demands of business? Let's look a little deeper. If we assume an antagonist can build a Deep Crack look-alike for your chosen algorithm, is prepared to invest say $27M dollars (128 times the previous budget), and improvements in technology give a four-fold increase in search speed of the machine, this all adds up to give an effective hit rate of $2^{47} \times 2^{2} \times 2^{7} = 2^{56}$ keys/sec. If we are using one of the more sophisticated of today's systems with a 128 bit key, Deep Crack II should find the right key (on average) in $2^{128-56}/2$ seconds, which is about 74 billion years. Perhaps the worries over net security are based more on perception than actual threat.

One thing that our little example does explain is why one common approach, triple DES with a longer key than DES, is expected to remain safe from attack for some years to come. Triple DES encrypts the source data three times using the basic DES algorithm, first using one key, then using a second key, then again with the first key. The effect is regarded as being equivalent to using a 112 bit key—long enough to be safe but short enough to be practical in terms of efficiency. Deep Crack II would still have to toil for nearly a million years to crack Triple DES.

There is no room for complacency, though, as computing power is continually increasing and there is a corresponding demand for ever longer key lengths that will protect against attack. Two main factors hold back the use of very long keys: first, the longer the key, the more processing power is required to perform the encryption and decryption processes (and hence the longer it takes for legitimate users to send secure messages), and secondly, as mentioned earlier, governments (particularly the US government) place restrictions on the export of (and in some cases the use of) strong cryptographic products. There are still some US legal limitations to the use of DES; 56 bit DES can be used by any customer, but US export restrictions limit Triple DES to financial services applications.

Successful use of encryption technology depends not only upon proper management of encryption keys (including changing them on a regular basis), but also on secure housing of the encryptors.

Ideally, the encryption should happen in the end system hosts and terminals so that there is no part of the end-to-end path left unprotected. This is not always possible, though, and a compromise solution is to encrypt data within specialist network devices (e.g. routers), or to place stand-alone encryption hardware at designated sites.

Most of the stand-alone encryption systems available on the market are designed to operate over point-to-point synchronous data links—but an increasing number of systems are being developed that are able to work over packet oriented networks, like the Internet. These packet oriented encryptors are also known as 'payload' encryptors, as they only encrypt user data, leaving protocol headers in the clear, so that packets can be properly sent across the network.

The latest standards aimed at providing secure, encrypted tunnels across the public Internet come from IPSec (a working group within the Internet Engineering Task Force defining a set of specifications authentication, integrity and confidentiality services at the IP datagram layer). Unlike DES, which generally operates at the application level of communication, the IPSec standards are integrated with the IP level of TCP/IP communication. There are basically two parts: encryption of the contents of IP packets (the IP payload encryptor, which uses a DES-type algorithm); and a packet header protocol (MD5/SHA), which provides added integrity checking. Key management is effected using RSA public key methods.

5.2 DIGITAL SIGNATURES

We have stated that public key cryptography, using public/private key pairs, provides a suitable technology for the creation of 'digital signatures' (Figure 5.3). If, for example, Steve encrypts a message using his own *private* key, and Mark is able to decrypt it using Steve's *public* key, Mark

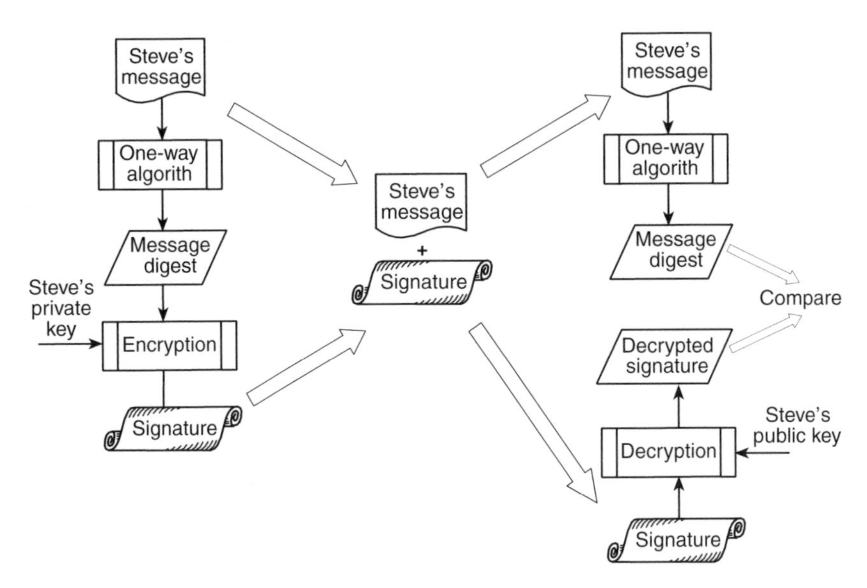

Figure 5.3 Digital signatures

can be highly confident that the message originated from Steve. If the message authorises Mark to perform some action (such as to take some money out of Steve's bank account), the fact the message was provably sent by Steve gives a similar level of confidence as if Steve had handwritten a signature on the message.

The technique described above is a bit unwieldy, with the 'signature' actually being the encrypted version of the entire message (which could be very lengthy). In reality, a digital signature is created from a *digest* of the message rather than the entire message. This digest is created by passing the entire message through a 'one way function' (also known as a 'hash' function), which has the following properties:

- it is 'one-way' in the sense that it is easy to convert the original message into a digest, but impossible to use the digest to recreate the original message;
- the digest is of a small and standard length, irrespective of the length of the original message;
- every character of the original message is significant in creating the digest, so that changing a single character in the original will result in a different digest;
- the function is sufficiently complicated that it is very difficult to make multiple changes to the original message so as to yield the same digest.

This last point is necessary to prevent the possibility of an unscrupulous person making some change to the message (such as adding an extra zero to a sum of money) and then making other 'compensatory changes' so that the entire message still gives the same digest as the original.

The algorithm commonly used for creating digital signatures (it is used, for example, in the SET banking standards) creates digests that are 20 bytes long, irrespective of the original message size. This algorithm meets the requirement to be one-way in nature, apart from anything else, reducing a long message down to a 20-byte digest must of necessity 'throw away' a lot of the original information, ensuring that there is no possibility of reverse-engineering the message from the digest.

Despite this apparent throwing away of information, the algorithm is such that changing a single bit in the message will change, on average, half of the bits in the message digest, and the probability of any two messages having the same digest is one in 10 to the power of 48 (a number bigger than the number of atoms in the solar system!). Hence, it is not computationally feasible to generate two different messages that have the same message digest.

So, if Steve wants to sign a message he is sending to Mark, the actual process is:

1. Steve creates his message.
2. Steve (or his software) creates the 20-byte digest of the message.
3. Steve encrypts the digest using his private key—the resulting object is the digital signature of the message.
4. Steve sends the message together with the digital signature to Mark.
5. On receiving the message, Mark also generates a digest of it.
6. Mark decrypts the signature using Steve's public key.
7. Mark compares the decrypted signature with his locally-generated digest.

If the two match, Mark can be confident that the message was signed by Steve and that it has not been tampered with by anyone else.

5.3 TRUST

When I access an Internet site that calls itself American Express, Ford or Barclays Bank, I want to be sure that it is truly what it purports to be before I engage in a financial transaction with them. The way this is done is to have the site certified by a 'trusted third party' who checks the authenticity of the site and provides certified copies of that site's 'public key' for communicating securely. VeriSign Inc. are world leaders in providing these services, and in the UK, BT is providing similar services (BT TrustWise) in partnership with VeriSign. We now look at the types of certificates that are issued by such trusted third parties, how trustworthy they really are, and what types of electronic transaction they are appropriate for.

The asymmetric encryption algorithms described above form the base of public key infrastructures. As indicated above, public key schemes are widely used for electronic document signatures (to prove source and provide non-repudiation of transactions) and for exchanging data encryption keys. A user will generate a public and private key pair, and publish the public key. However, we do need to know that the public key belongs to the right person and not an impostor. The public key infrastructure achieves this by having a trusted third party who signs the users private key, after the user has proved their identity to the third party. This signed key is called a certificate, and they are issued by a Certificate Authority (CA) such as VeriSign. The certificates need to be accessed from a directory, so that a message recipient can retrieve reliable public keys.

A certificate itself is basically a set of data elements, bound together and electronically signed by a trusted CA using its closely guarded private key, as illustrated in Figure 5.4 The Figure shows the structure of a certificate that follows the X509 standard format. The top portion of the certificate comprises straightforward factual information (that anyone could reproduce). The important feature is the signature the end of the certificate.

Figure 5.5 shows the process that has to be followed for this essential part of the certificate to be appended. A digest is formed from the top

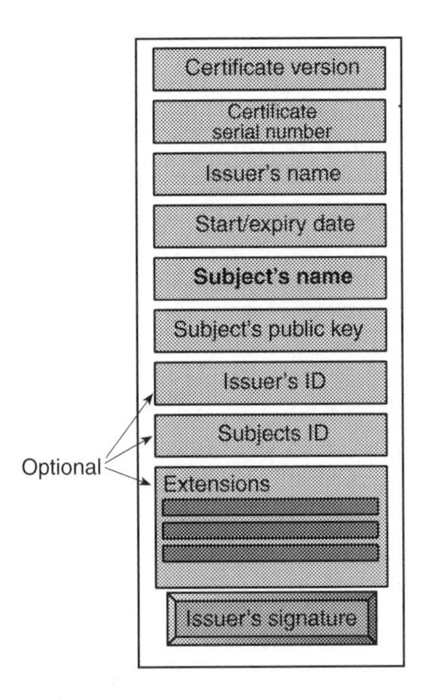

Figure 5.4 A digital certificate

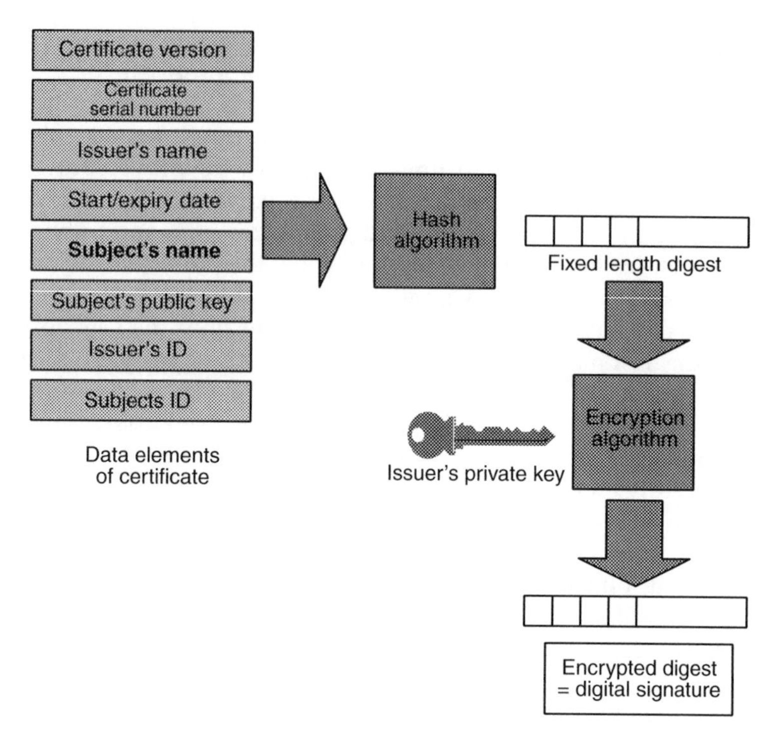

Figure 5.5 Signing a certificate

portion of the certificate, and this is encrypted with the private key of a trusted CA, such as VeriSign.

The certificate's authenticity can readily be checked by comparing the digest generated from the top half of the certificate with the digest obtained from decrypting the signature with the issuer's public key (Figure 5.6).

There are several classes of certificate. Each has a different use and calls for different data elements to be incorporated in the top half of the certificate. A basic Class 1 certificate binds a user's name with their e-mail address and their public key. This is used by individual internet users to send secure e-mail or to identify themselves to web servers.

A Class 2 certificate is issued by an organisation such as a bank, to identify its customers. It has bound into it further details such as account numbers. The certificates are still issued by a CA, but the users' applications are processed by an in-house RA (Registration Authority), which is used to approve requests for certificates (e.g. once credit checks and written application for service is received). BT TrustWise Onsite is one of the UK providers of an RA facility.

Web server operators will seek a Class 3 certificate. In this instance, the CA will carry out rigorous checks to confirm the identity of the server owner and that they are a creditable organisation. The certificate binds the

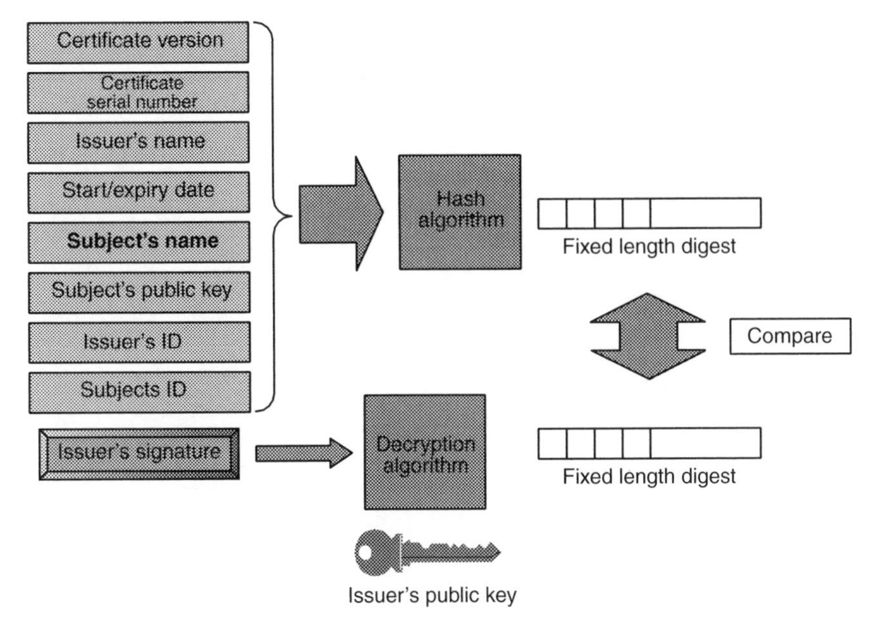

Figure 5.6 Validating a certificate

server's URL, the organisation name and its public key. If a server has such a certificate, browser users can confirm they are communicating with a genuine server and not an impostor. More importantly, the certificate enables encryption keys to be exchanged and secure HTTP sessions to be run (e.g. for e-commerce and electronic banking applications). For example, you would use the public key from the certificate to set up an encrypted link to the merchant's server. A symmetric session key (ordinary private key) and the public key from the certificate would both be used for encryption during the session; processor intensive PKI to transfer the small amount of data needed for key exchanges, and bulk transfers using the more efficient private key method.

Of course, the use of a Certification Authority only serves any purpose if they are indeed a trusted third party and that they genuinely carry out adequate validations of the data that they certify. So, how do we know that we can truly trust our trusted third party? Are 'Certificates R Us' a trustworthy purveyor of digital certificates? The answer is that the public key of the CA is itself distributed with a certificate issued by an even more highly trusted institution (a trusted fourth party?). Thus, we can end up with a hierarchy of trust, in which each party is certificated by another higher up the chain. At the very top of this hierarchy is an authority such as the RSA, who will only certify other parties following an in-depth investigation of their security credentials. This hierarchy of trust is illustrated in Figure 5.7.

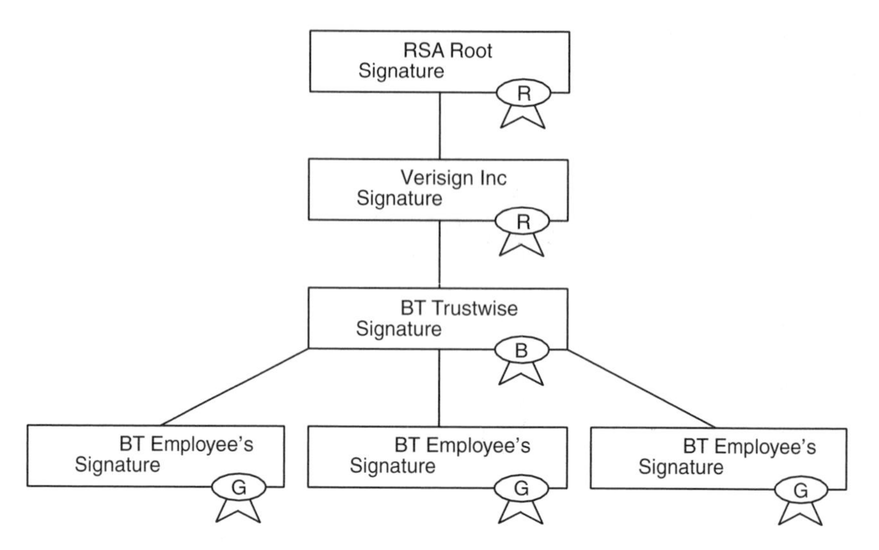

Figure 5.7 The hierarchy of trust

One important design issue with PKI is where the private keys should be stored. If on a PC, the key only really proves a message came from the PC, not the person. Simple user passwords are used to protect the private key on the PC, and this offers only weak protection. A much better option is to keep the private keys on a smart card that moves from machine to machine with the key owner.

When small secrets (keys) are used to protect big secrets (data), this is an altogether better option—we now take a brief look at smart cards.

5.4 SMART CARDS

Currently, many users of public key cryptography schemes store private keys in software (e.g. using encrypted directories on their hard disks). As we have said, this is far from ideal as the private key is only protected by a single password, and there is little support for mobility.

Smart cards provide a convenient, mobile alternative for key storage. Security of the system is increased, since access to the private key is now controlled by something the user owns, as well as something the user knows (the password). The similarities between smart cards and existing credit/debit cards also ensure that many users are already familiar with the basic principles of card usage and management.

Current barriers to widespread adoption are largely associated with the lack of agreed smart card standards (e.g. Microsoft and Netscape support different standards for integrating smart cards with their web browsers) and the corresponding lack of smart card 'slots' in PCs. Recently, a vendor has produced a low-cost replacement PC keyboard which incorporates a smart card slot. Another vendor (Aladdin) produces a dongle which is functionally equivalent to a smart card but which plugs into any USB socket on a PC. Yet another vendor produces a card reader which can be plugged into a floppy disk drive. However, on the question of which, if any, of these approaches will win in the market, the jury is still out.

Key management is by no means the only use for smart cards within e-commerce. Smart cards are increasingly being used to support a wide range of applications, including user authentication, electronic cash systems and loyalty schemes. With the advent of multi-application smart cards, it is expected that several of these functions will be combined onto a single card. For example, financial services companies may issue a single smart card combining traditional credit/debit card facilities with an electronic purse, strong authentication for home banking and key management to support secure electronic payments.

5.5 SECURITY GOOD PRACTICE

The previous sections have concentrated on the specifics of how to provide a secure environment for electronic transactions across the Internet. There are, of course, many other security considerations to be borne in mind, but most of these are not specific to eBusiness. For example:

- Physical server security (where are your server computers housed, and who do you trust to feed and water them?). Discussions on eBusiness security tend to centre on the network and application security issues. However, in practice, most security breaches are perpetrated by companies' employees rather than by Internet 'hackers'. The main things to consider are the physical location of the servers (and who has access to them), and the processes for managing them. What, for example, is the process for disposing of backup tapes, and who is responsible for regularly changing the root password of the server?
- Firewalls and proxies (how do you stop people hacking into your servers?). When providing servers attached to the Internet, it is customary to control access via a firewall machine. The rationale for this is even more compelling if the servers contain key business and financial information. Essentially, the firewall needs to restrict access to just those facilities that you wish to offer to your customers (it might, for example, bar all protocols except HTTP and secure HTTP).
- Denial of service (how do you stop someone flooding you with bogus transactions that prevent your real customers from using your services?).

One of the reviled practices on the Internet, known as 'spamming' is to send unsolicited mail to large number of users on a frequent basis. Address screening may be required as some incoming traffic may have to be barred.

- Auditability (how can you prove that the transactions you claim took place really did so?). Most network equipment keeps an extensive record of who called, what they did and when they did it. It may be worth storing these records somewhere handy.
- Embedded security (are specific applications applying precautions?). Most people will have noticed the open/closed padlock or broken/whole key in their browser. Applications such as Internet Explorer and Netscape Navigator can invoke a secure session by invoking a link using HTTPS, instead of HTTP. This calls the Secure Sockets Layer (SSL), which initiates a session to the server using port 443 instead of port 80 (the well known port for HTTP). SSL then works with TCP to set up a secure session.

As well as being fairly general, the above issues are predominantly the concern of a network/service supplier. Others would want to know that they have been attended to, but would probably not have to take direct action.

5.6 PUBLIC KEY INFRASTRUCTURES (PKI)

We have seen that the use of public/private pairs of encryption keys provide a valuable mechanism for two parties to trust each other. The emphasis so far has been on how we distribute and certify the keys. This, however, leaves some other questions unanswered: how do we revoke a certificate from someone we no longer trust? How do we create blacklists of untrusted people and companies? What can we do for someone who loses their private key? How does a company that has implemented encryption gain access to data that has been 'locked-up' by an employee who is no longer trusted (or has died or left the company)? All these and other issues are addressed by the implementation of a Public Key Infrastructure that manages the keys for a company or other organisation.

The basic job of a CA is to issue digital certificates. If Mark purchases a personal (Class 1) certificate from, for example, BT TrustWise, he will fill in an online form with some personal details about himself (including his email address). Then, behind the scenes, his web browser will create the public/private key pair for him. The private key is stored locally (either as a password-protected file on his disk or perhaps on a smart card), and the public key is sent to the TrustWise CA. The TrustWise service creates the certificate which includes Mark's name, email address, public key and other information, then it signs this certificate using its own private key. It forwards the resulting certificate to Mark via an email message.

For Mark to make use of this certificate, he must give copies to his friends and acquaintances so that they can use it to encrypt messages that they send him: he can do this by emailing copies of the certificate which are then stored by the recipients' email systems. However, there are limits to the usefulness of this approach: having given Mark his certificate, there is no easy way to take it away from him if we cease to trust him, and also the above approach is only practical if Mark has a manageably small number of acquaintances who want to send him secure emails.

In the case of a large company, with thousands of employees who regularly send each other confidential information, the above approach is inadequate, so the first step in creating a public key infrastructure is to create a corporate email directory which includes digital certificates for employees. If one of these employees wishes to send mail to another, they will first look up the recipient's email details in the directory and will also (probably behind the scenes) take a copy of the recipient's digital certificate. Hence, we now have a much more scaleable solution for distributing certificates. Also, the directory is the central point at which certificates can be allocated to people in the first place, and it provides the place from which certificates can be removed or revoked. This latter function only works if email clients check with directory every time a message is to be sent: in practice, they are likely to cache the certificate, so currently there is not a completely satisfactory solution to the problem of certificate revocation.

The usual means of creating the public/private key pair is to do this at the user's browser, as described above. Therefore, the private key never needs to be sent over the network or, indeed, to leave the user's direct control. However, this arrangement is not always appropriate to a corporate user of a public key infrastructure. Just as companies may insist that employees deposit spare copies of office and desk keys with their security department, they may want to keep copies of users' private encryption keys. This is required to protect against such eventualities as an employee leaving the company (or dying) suddenly, and for someone else having to deal with the encrypted contents of their email boxes. Similarly, an employee might lose a private key (e.g. through a disk crash) and again be unable to access stored messages. The answer to this is for the company to retain copies of all private keys in highly secure storage: this process being described as 'key escrow'. Some governments are also contemplating the use of key escrow as a condition for the licensing of strong encryption technology (i.e. you can only use strong cryptography if you lodge a copy of the key with a government agency). The social and constitutional issues around key escrow are beyond the scope of this book, but the main practical argument against it is that by storing copies of all private keys in a single location, one creates a very obvious target for hacking and espionage.

5.7 SUMMARY

Concerns over the security of the Internet, much hyped by the media, have done little to reassure users that the Internet is a safe place to trade. Indeed, security is cited as the number one barrier to the growth of online trade. If progress is to be made and the promise of eBusiness is to materialise, these concerns must be addressed.

This chapter has explained the various forms of attack that a net-borne business can suffer. It then goes on to detail the various mechanisms that are available to ensure security, privacy and trust over the net. The major security issues include:

- *Confidentiality*—data must not be visible to eavesdroppers.
- *Authentication*—communicating parties must be certain of each other's identity and/or credentials.
- *Integrity*—communicating parties must know when data has been tampered with.
- *Non-repudiation*—it must be possible to prove that a transaction has taken place.

In terms of solutions, we explain

- *encryption*—for data protection and assurance.
- *digital certificates*—that allow trading parties to know who they are dealing with,

and how a public key infrastructure, along with strong encryption algorithms, make these security measures a practical proposition.

FURTHER READING

Garfinkel S & Spafford G, *Web Security & Commerce*, O'Reilly, June 1997.

OECD, *The Economic and Social Impact of Electronic Commerce: Preliminary Findings and Research Agenda*, October 1998 (http://www.umich.edu/~dirsvcs/ldap/doc/rfc/rfc1777.txt)

ITU Recommendation X.509 (11/93), *Information Technology—Open Systems Interconnection—The directory: Authentication framework*. ITU, Geneva, 1993.

Secure Electronic Commerce Statement, Department of Trade and Industry, 27th April 1998 (http://www.dti.gov.uk/public/frame9.html)

Secure sockets, IETF specification and vendor implementations (http://www.phaos.com/sslresource.html)

6

Integration

The golden rule is that there are no golden rules

George Bernard Shaw

Integration has never been one of the sexy topics in computing. Yet, like the poor, it will always be with us and will always need care and attention. Integration is particularly relevant in eBusiness as it is the means by which the real magic of online trade is delivered. Looking back at the market models introduced in Chapter 2, it should be fairly clear that they only really work if they have all the information they need, and quickly too. Much of this information will reside on a database that was installed ten years ago, legacy (aka cherished) inventory records or a third party customer handling system. Somehow, disparate and distributed systems have to be made to work together and to deliver what is needed, when it is needed.

Because it is largely invisible to the user, integration lies at the other end of the appeal spectrum from the glossy, shop-window side of eBusiness. However, just because integration isn't something new and isn't particularly visible to the end user, it does not mean that it isn't important—it certainly is. It delivers the engine that allows the on-screen promises to be realised.

To some people, integration is achieved as soon as a number of components have been connected—rather like welding together two halves of a car. The result is about as satisfactory in both cases. So, in this chapter we will assume that the job is not complete until there is seamless interworking between all of the various parts of a proposed eBusiness solution. After some definitions and general integration principles, we go through two case studies to show what can happen in practice. To close, some of the strategies and tactics for integration are explained.

6.1 DEFINITIONS

It would be nice if we could define exactly what is involved in integration. It would be even nicer if this definition indicated that a well developed set of techniques were available to deal with it. In truth this would be to give a false impression—integration is often seen as a necessary evil rather than a soundly based aspect of engineering. So, our definition is that back-end integration should cover all aspects of getting the various components of an eBusiness solution to work together so that it is fit for purpose. In practical terms, this is just about as difficult as a complex mathematical integration—and considerably less well defined.

From our definition, we are concerned with the way in which two (or more) systems are made to communicate with each other. At least one of these systems will be part of an organisation's existing established systems estate—typically, an operational support system that holds some useful data on customers, products, service, etc. In the eBusiness context, integration entails, for example, the establishment of a communication link between an online server and a merchant's back office product database. This 'back-end' integration often presents a real challenge, and we'll examine what the issues are in our case studies.

The systems being integrated into an eBusiness solution tend to come in two main varieties, batch-oriented and real-time. Despite doing much the same thing, these two vary greatly in the functionality they provide, and in their complexity to implement.

For our purposes, real-time describes the ability of an eBusiness system to interact with another system during the order process. This might be to obtain data that is too volatile to store in an online system, to retrieve data that is too sensitive to store in the online system, or update data in one of the systems based on an event. While this interaction must be fast, delays of a few seconds are quite acceptable. So our definition is not quite as stringent as the traditional one, where delays are no more than fractions of a second.

By way of contrast, batch-oriented denotes the common practice of sending data between two or more machines at set intervals. Data that needs to be sent to update information on another machine is gathered together and held locally until the next scheduled transfer time. Thus, an eBusiness server using a batch-oriented interface would stores purchase orders locally, and, at set intervals, forwards the orders to the back-end system. A similar approach would be used to transfer price, product and stock inventory updates from the back-end to the eBusiness server. Batch-oriented back office integration normally uses structured text files as the medium for passing data.

Each approach has its merits and limitations. These will be explored in detail in later sections.

6.2 INTEGRATION PRINCIPLES

Perhaps the first thing to say is that integration can be something of a messy affair. This shouldn't be too much of a surprise, as it calls for the seamless union of at least two things that were never originally intended to be coupled. The case studies presented later on more than adequately illustrate this. Despite there not being a prescription or formula for back-end integration, there are some preferred technical options, and that is what we cover in this section.

Let's start with a fairly simple example. A merchant wants to enable customers to be able to check the status of their accounts. This merchant already has an established web site, but all of the relevant information needed for this new service resides on a database in the merchant's head office. So all of the elements exist. It is simply that they do not work together to provide what is wanted—a little work is required to supply the service in a usable way.

There are some quite simple solutions that can be applied. The most straightforward is to use a scripting language such as Perl or CGI to retrieve records from the back-end system. In this instance, a request from a client (i.e. an entry on the merchant's page via the browser on the user's PC) would be passed from the server hosting the merchant's web pages to the back-end system via a Common Gateway Interface (or CGI) script. What this script does is

- to reformat information received from the customer
- to pass it on (in this instance as a database query) to the back-end system
- to retrieve results and to present the requested information to the user through a browser as if it were a sourced from the local server.

This is illustrated in figure 6.1.

Typically, scripts would sit on the merchant's server—in practice they are no more than a few files (usually located in the /cgi-bin directory). Although fairly straightforward to develop, scripts are more often bought in. For instance, several companies offer scripts that allow a 'buy' button to be put on a web page and which take care of all of the back-end payment and settlement transactions.

Scripts provide a quick, low cost solution to some integration problems. However they do tend to be limited to the more straightforward applications. When more complex (i.e. diverse or high transactions volume) applications are required, a more engineered approach is usually called for. It is here that distributed system technology (such as CORBA) can be applied to ensure that several systems can co-operate on a task, such as the assembly of a customers purchase history where products, prices and customer

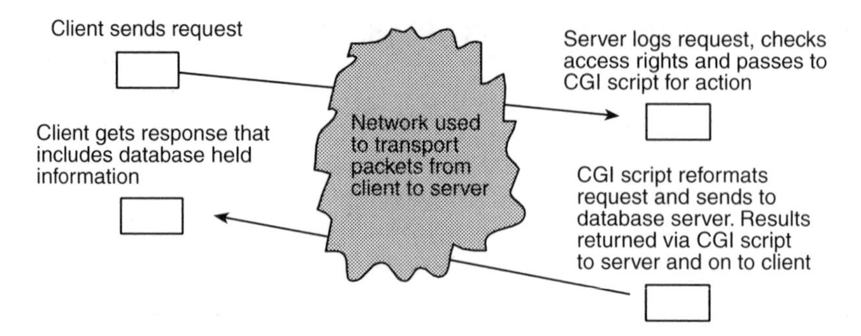

Figure 6.1 Back-end integration using a script

records are held on different back-end systems. The other motivation to move towards this option is to provide scalability—many online services are now recording phenomenal hit rates, and commercial-strength technology is needed to accommodate this.

Some of the relevant technology for distributed systems is explained in Chapter 9. We will not delve further here, as it is a mainstream area of computing and is, therefore, well supported with books, suppliers and know-how.

6.3 PLANNING INTEGRATION

Integration doesn't just happen. There is some preparatory work that needs to be done up front, and then a process that needs to be followed to complete the job. This is really no more than good engineering practice, and so follows a well-trodden path. So, before anything else, it is important to be clear in exactly what needs to be done, what approach to take and how to go about the task.

Information

There are two connotations to the word 'information' here. The first is the information that needs to be sought from the back-end system. Some of the questions that need to be asked early on include—exactly what data is needed, in what format, how frequently and with what accuracy? It is well worth producing detailed message flows (rather like those used in Chapter 2 to describe the various trading models) to show what goes where and in what order. Before committing any time or money, there should be a very clear idea of why the back-end system needs to be integrated, and this should be supported with a detailed picture of how it interfaces into the eBusiness process it will be supporting.

The second form of information is technical information on the back-end system itself—its interfaces, data formats, etc. In much the same way that is easier (usually) to deal with someone familiar, so it is more straightforward to integrate a system that you know something about. There are at least two problems with established familiarity, though. First, the job of maintaining a back office system does not call for superb communications skills. Information can be difficult to extract from those who really know the how system that is to be integrated actually works. Secondly, the back-end may be provided by a third party, so they may be less than willing to share what they know. Given that detailed technical knowledge is often important, this can be a significant practical issue.

Type of interaction

In our definition of integration we mentioned two basic styles—real-time and batch. It is important to be clear in which style is appropriate. The wrong choice can lead to unnecessary cost or complexity (where real-time is used and batch would do), or to inadequate functionality (where batch is used but real-time was needed).

As a general rule, batch should be the default, and should be used whenever a customer does not have specific and real requirements that can only be met using a real-time interface. The simple reason for this is that a batch-oriented interface is quicker, easier and less costly to implement.

Real-time integration is difficult to implement, and should only be used when there is no alternative. Most of the instances where this would be the case are to do with money:

- Dealing with the volatile stock levels of their inventory. Quotations that are even a couple minutes old may be too stale to be useful.
- Performing real-time credit checks. An organisation may not want to allow someone to make a purchase on faith that they have the credit to cover it. Credit information on customer accounts is considered too sensitive by many companies to allow it to reside on an outside server.
- Provision of time sensitive information such as stock market quotes. Such premium services would have to perform.
- Providing current account and order status. Organisations can save money if their customers use the catalogue for information they would normally call a customer representative for.

- Running credit authorisations through the back office. Some organisations may prefer that credit cards be authorised through their own acquirer via their back office accounting system instead of using an outside system.
- Displaying prices that are based on volatile market rates. Organisations that deal with commodity items or extend loans may tie their rates to market prices at the moment of sale.
- Matching services provided by a competitor. Customers may start using a competitor's catalogue if it has more functionality.

The style of interface is largely independent of the first preparation step for integration—the definition of message and data flows. It does, however, have significant impact on the way in which the integration is carried. For this reason, a case study for each style is given.

Process and practice

Integration must be carried out systematically. By its very nature, there is an element of trial and error in every integration project—few legacy systems come with comprehensive documentation. Hence, it is important that specific sets of circumstances can be recreated and that a given system state can be recreated.

There are two key aspects to integration process and practice. The first is to have an established way of doing the job that has defined stages, each with specific inputs, outputs and transformations (Norris, Davis and Pengelly 1999). The quality systems used to run software development projects exemplify the approach and standards such as ISO 9001 give a framework for a good integration process. A second key element is configuration management, which entails keeping control over each component part of the overall system and knowing how they are combined. As well as knowing what the history of each part is (i.e. what has changed and which version is used where), it is important to have a known baseline that can be returned to if all else fails. The other essential part of good configuration control is a compatibility list that defines which version or variant of each component works with which others.

Once all of the mechanics are in place and the nature of the integration task is understood, the map is complete and work can begin. Of course, the map and the terrain don't always agree, so we now go on to look at a couple of real world examples.

6.4 BATCH-ORIENTED INTEGRATION

In this case study, a medium sized manufacturer of decorating supplies wanted an Internet catalogue that they could use to provide product and promotional information to their resellers, and also to accept product orders online placed through the catalogue

Given that this project predates our 'batch as default' guideline, the choice of batch-oriented integration was made for two reasons. First, supplier inexperience in catalogue development made the complications of real-time integration look too risky. Secondly, the merchant had no requirements that made real-time integration necessary. In fact, their IT infrastructure and inventory tracking system were not sophisticated enough to provide accurate, real-time data about inventory levels and customer orders.

In this case study, the back-end integration requirement was for a system to send files between the merchant's catalogue and their back-office. The files from the catalogue consisted of purchase order requests from the merchant's customers, while their back-end sent files containing batch updates of product pricing and order status notifications. A simplified illustration of the message flows is shown in Figure 6.2.

This is not very different from the example given earlier on for retrieving records from a back-end database using CGI scripts. Indeed, the technical approach adopted was much the same, although this is not where most of the real issues lay. The actual implementation threw up a number of practical points and, with the devil in the detail, here are a few of the more interesting ones.

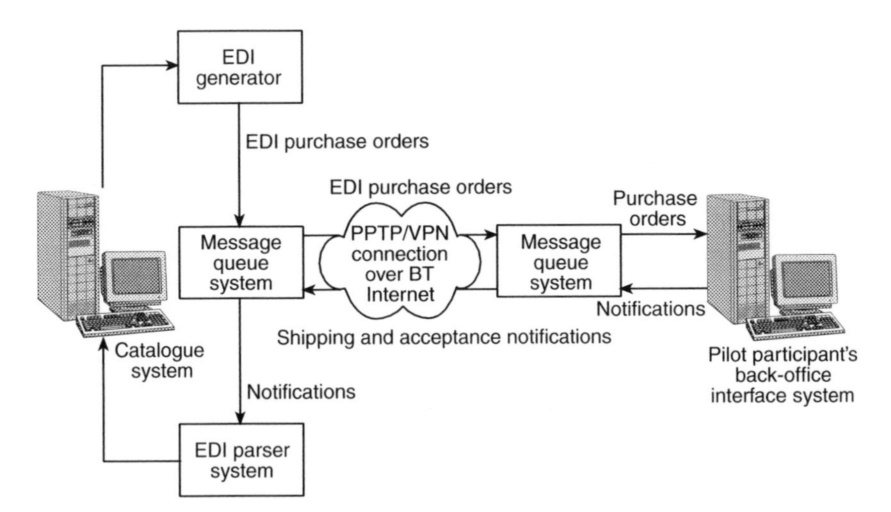

Figure 6.2 Batch-oriented integration

Standards are not always standard

With an eye on reusing solutions in subsequent production systems, the EDIFact standard was adopted for purchase order messages sent from the catalogue system to the pilot participant's back office. On the face of it, the messages defined in the standard provide a set of strict definitions, and this ensures that a basic purchase order from one company would be easily read and understood by any supplier adhering to the same standard. While EDIFact messages do contain a structured format, many of the message definition areas are vague, and these areas must be negotiated between the two parties exchanging the messages.

The EDIFact standard also gives many optional fields that may be used to express discretionary information about an order. Furthermore, some of these optional fields have almost the exact same meaning. Thus, trading partners must negotiate which fields within a specific message type will be used.

On the implementation side, the standard describes what message formats and fields a parser needs to handle in order for it to be EDIFact compliant. While the fields are strictly defined by the standard in the length, type of characters they can contain, and the order they appear in, only vague guidelines are given as to what the actual content of a field should be. Hence, even when the field content is well defined, the meaning of that content is not. For example, one field that has different meanings is the quantity field. So let's say that one supplier sells beer in packs of six, and so the item is a six-pack of beer. The supplier may get purchase orders from two buyers, one requesting a quantity of one six-pack and the other requesting a quantity of six beers. While the number in the quantity field is different, both buyers are requesting the same thing—a six-pack.

With some buyers using the quantity field to state the number of packs of an item they wish to purchase, and others using it to designate the number of individual units of an item they wish to buy, there is inevitable confusion. Thus, the supplier has to use prior knowledge about the buyer's message format to determine if the quantity six means half-a-dozen six-packs or six individual units, i.e. one six-pack. Regrettably, there is no 'we are having a party' field to help the recipient decide which option to choose.

The obvious candidate is not always the right one

Having decided to implement the batch-oriented data transfer approach, some method of transporting data between the two machines needed to be chosen. An obvious candidate was SMTP (the Internet's Simple Message

Transfer Protocol) email, a standard protocol. With all eBusiness participants having a connection to the Internet, it is reasonable to assume that they can all receive SMTP email. In addition, it would appear to be an easy, reusable solution.

Further exploration of using email as a transport mechanism leads to doubt, though. Several factors about SMTP *per se* and about specific email implementations give some cause for concern:

- SMTP compliant email systems are not as ubiquitous as they seem. Most corporations use proprietary LAN-based email systems such as MSMail and Outlook for sending internal email, and use a gateway for sending and receiving SMTP mail.
- Since LAN-based email systems and their gateways are proprietary, any solution tailored around the one merchant's mail system could not readily be reused with an organisation that uses a different email system.
- SMTP gateways tend to be attached to the normal TCP/IP port for Internet mail, so it is not possible to install a specialised SMTP mailer on that port.
- SMTP mail does not have a reliable delivery notification mechanism, and many applications need a 100% failsafe approach.

So, if the obvious (ubiquitous and open) candidate for message transport poses so many problems when it comes to implementation, what next? If the standard option is wanting, maybe a proprietary answer would work, which brings us to the next lesson learned.

Proprietary can be good, in a closed environment

While SMTP mail is an obvious choice data transport method, Microsoft's PPTP protocol probably isn't. PPTP is a proprietary protocol (albeit proposed as an open standard), so it only currently only runs on Microsoft platforms. Most good system engineers are disinclined to adopt proprietary protocols, as they tend to inhibit flexibility and lock you in to one supplier.

While PPTP is a proprietary protocol, there are many occasions when the back-end is a closed environment. If the back-end system(s) do not have to talk to a variety of machines from the outside world, a proprietary option makes sense. In this case study, PPTP was only used to communicate between two NT systems: the electronic catalogue and the merchant's electronic catalogue-related messaging workstation.

So, open protocols may be a nice, elegant strategic goal, but there are many situations when a proprietary solution is more appropriate tactical solution. Back-end integration is often concerned with closed systems, and these are more suited for proprietary solutions (or, at least, they benefit less from an open solution). In the fast moving world of eBusiness, it is important to not let developer prejudice or idealism stop people from looking at what may be the best solution!

Control is good

One fundamental assumption that proves to be true over and again is that every company has a different back-end configuration. This poses the challenge of how a supplier can create a reusable, batch-oriented interface, when each organisation they provide a solution for has a different back-end environment.

Perhaps the most workable solution in this instance is to assume control of both the hardware and software at the catalogue server and back-end interface. This ensures some measure of control and commonality, as there is at least one machine on each merchant's network that can be communicated with. The third party supplier can require each merchant to install the same communications software on a dedicated workstation.

6.5 REAL-TIME INTEGRATION

In this second case study we have an electronic parts distributor who see their competition moving from CD-ROM based catalogues to Internet catalogues for providing product information and allowing customers to place orders. They needed to keep up with their competitors, and also saw the Internet as providing a new channel to market.

In this instance there were some very clearly articulated requirements for a real-time interface to the merchant's back office. First, the dynamic nature of their inventory stock levels required the electronic catalogue to be able to allocate and deallocate inventory from the back-end systems during the order process. In addition to this, the merchant trading model required a customer to be told if an item is out of stock prior to the order being finalised. It also required that the order entry system confirm that the customer has avan ailable credit limit to draw on. The credit checks review both the customer's outstanding balance and any unbilled back orders.

So there was little doubt in this instance that information had to be shipped to and from the merchant's back-end systems within a few seconds of the customer completing a transaction on the merchant's web pages. The outline flow of information in this instance is illustrated in Figure 6.3.

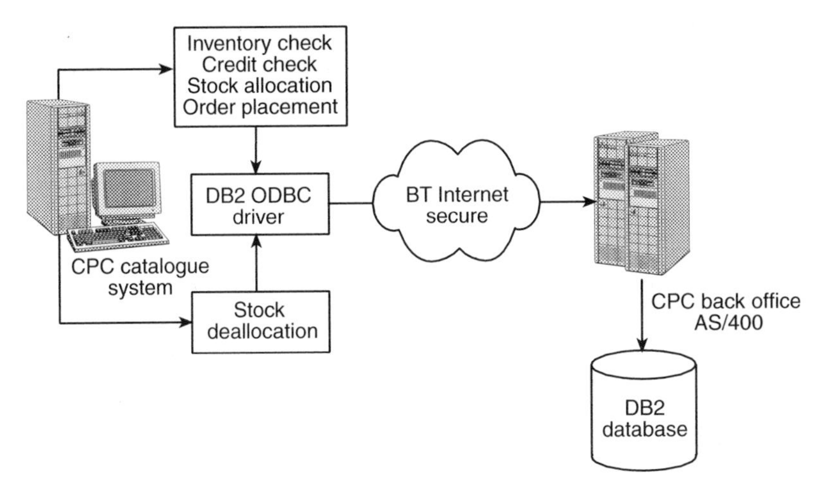

Figure 6.3 Real-time integration

Because of the need for credit checking, the link to the back-end systems had to be protected, hence the use of a secure network. That aside, the picture is not radically different from the previous batch integration solution. Given this, there were certainly common lessons related to requirements elicitation and information gathering. However, the issues reported below are those specific to the creation of a real-time back-end solution for a merchant.

Clean that data before you bring it in here

One of the perennial problems of integration, especially with legacy and back-end systems, is that of 'dirty' data. In this instance, the merchant's production system provided product pricing and other data to the electronic catalogue. This data was not directly useable, though, so the extra task of data cleanse had to be inserted into the development plans. In retrospect, it should probably have been there to start with.

The problem on this occasion was that data had to be transferred between different types of databases. All of the merchant's product maintenance data was held in an AS/400 DB2 environment. It may very well have been clean data, but when transferred to the Oracle database that formed part of the eBusiness system there were problems. The Oracle database could not correctly handle the data from the AS/400, as many entries in the DB2 database had leading or trailing whitespace, and it was impossible for the Oracle database to match it with already existing data.

The solution to the whitespace problem with the data was to filter out the extra whitespace from all data received from the back-end. While filtering white space may not be an issue for all implementations, experience suggests that it is prudent for developers to allow for the implementation of filtering code as a standard part of any back-end integration plan.

By the way, it has to be 24 X 7?

One of the big selling points of eBusiness is that it effectively extends the hours that a merchant is open for business. The web is available 24 hours a day, seven days a week, 365 days a year (and more every leap year), so an online catalogue needs to be available all of that time.

One of the implications of integrating back-end systems is that they become an extension of the catalogue and are, therefore, subject to many of the same requirements. This raises an interesting point. Traditionally, many back-end systems have to go off-line at regular intervals. It is standard practice for some to go into a batch-processing mode each evening and for others to be taken down for routine maintenance at regular intervals. During this time, they (and the data they hold) are simply not accessible. Hence, there is an operational mismatch between the front and back ends of our eBusiness solution.

Some compromise is clearly needed and, given that it is unlikely that established technology will change, the fix has to be an operational workaround. Most merchants will be adamant that their catalogue is available all of the time, and they will want their customers to be able to log on and order at any time. Some scope has to be built into that allow users to select but not finalise their orders while the back-end systems are unavailable. This my well be a simple piece of exception handling—but some explicit design is required to prevent orders getting lost or applications freezing for want of input.

Testing—1, 2, 3

A controlled development environment is a wonderful place to acquire a false sense of security about the stability of a system, and how smoothly it will interact with the merchant's back-office once it goes live. Unfortunately, this false sense of security must be dispelled, and preferably before the system goes live. This is where acceptance testing plays a role.

Good software development practices dictate several levels of testing before any system is released. Unit testing is done first to find problems within individual components of the system. System integration testing is designed to pinpoint problems with the system as a whole. These tests do fulfil their purpose of catching most problems before the system goes operational. However, since the tests are executed in a controlled environment and are not subject to the unexpected events that are occur in a production or live environment, complete with real people, not all problems are found.

Customers who use (and abuse) an 'integrated solution' have an uncanny ability to uncover problems. Acceptance testing puts integration software through its paces, and makes sure that it both meets user needs, and that it integrates successfully with the back-end environment. In this case study, a three-phase test approach was adopted for the following reasons:

- It allowed a different aspect of the system to be tested in each phase, thus making it easier to find and correct problems that arose.
- It was possible to gradually move the system from a highly controlled environment to the actual production environment. Variables introduced by new network configurations and running over the Internet were slowly introduced during testing. This again made it easier to identify and fix problems as they arose.
- The first two testing phases established the confidence in the system before it was connected to the live back-end systems in the last phase.

The three phases were structured as follows:

- The first phase tested functionality of the new system. It did not involve connecting to the back end systems, but made use of a development machine (an IBM AS/400 that simulated the real back-end environment).
- The second phase tested the back-end integration logic by connecting to a test AS/400. This provided a controlled environment on a live site, where there was visibility of all the back-end database transactions. This allowed the capture of problems with back-end integration routines, but did not endangering the live AS/400.
- The final phase was to test the system against the live AS/400 before the site was announced to its waiting audience.

Mining for statistics

In the opening chapter, we mentioned that one benefit that online catalogues provide over paper or CD-ROM catalogues is that they can capture significant amounts of information regarding a user's interaction with the catalogue. A properly designed online catalogue will allow data mining for statistics beyond the normal user hits and products purchased. It allows a company to collect information such as which products users often look at but do not purchase, which categories of products are the most and least viewed in the catalogue, and what items users are searching for that may not appear in the catalogue.

The first step in designing a catalogue that maximises its data mining potential is to establish what data about customer catalogue viewing and ordering behaviour will aide the merchant's marketing efforts. What information do you want to learn about the users of the online catalogue? Is it worthwhile profiling all the products looked at by a customer before they decide on what to purchase? You may want to see what products are being viewed by a large number of users but not being purchased, and then create a promotional scheme around that product to see if it stimulates sales. Alternatively, you may want to create personalised mailings to users based on their buying and browsing habits. Once the data mining requirements are known, they can drive the design of catalogue data mining enhancements.

Site usage reports contain many useful statistics including bandwidth usage, site hits and user sessions. Each of these items can be useful and can be translated into decent market information. For instance, the number of user sessions (rather than site hits) gives the best picture of how many people are accessing the system.

One of the opportunities missed during this case study was omission of specific data mining requirements. Hence, data mining was not taken into account when the system design was finalised. The sting in the tail of this lesson is that with the system now live, there is valuable information to retrieve about catalogue usage that cannot be accessed, simply because the system was not designed to readily capture this data.

6.6 INTEGRATION TACTICS

Our final case study aims to provide a bit more detail on the tactical integration of back-end systems. In this instance, we have a productised Internet payment gateway that handles credit card transactions for online merchants. It provides a real time validation service for the main credit and debit cards (Visa, Delta, etc.). The merchant can access the gateway (and deal with the key eBusiness issues of authentication, authorisation and settlement) by installing a set of CGI scripts and a security programme on their server.

The logical layout of the merchant's web site, once the scripts are installed, is shown in Figure 6.4. The third party product is integrated with the merchant's existing online shop simply by configuring the new scripts. Once installed, the scripts allow a 'buy' button placed on a web page beside a product description to be backed up with online credit check, settlement and transaction logging. From the user's perspective, the back-end finance systems (which actually reside in an acquiring bank) have been

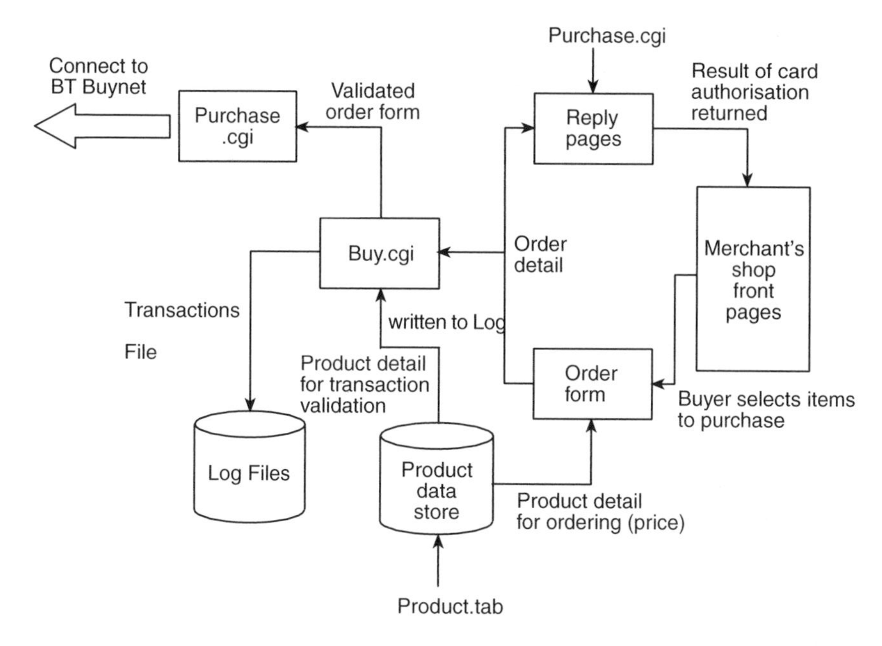

Figure 6.4 Logical layout of a merchant's site with integrated purchasing package

fully integrated. The main elements of the integrated system shown in Figure 6.4 are:

- *Catalogue pages*—this is the merchant's shop front. The look, feel and content of these pages is entirely under the merchant's control. The only effect of adding the new package has been to allow a 'buy' button to be added (by adding a couple of lines to the existing HTML).
- *Order Form*—a pre-configured order form is supplied as part of the product. All buyer details are captured in this form when an item is to be purchased. When the 'buy' button is selected, the details on the form are passed to the 'purchase.cgi' script, which validates the order details and initiates the authorisation process with payment gateway.
- *Product data store*—this contains information on the pricing of products for sale through the merchant's site. The 'buy.cgi' script (and hence the order form) uses the product data store to determine the price of the product.
- *Log files*—these record interaction with the payment gateway. The log files keep track of what was sold, to whom and when.
- *Reply pages*—these are the transaction status reports that return the result of a credit card transaction to a customer. Several default pages

are supplied with the product, and a custom page can be selected based on the return code sent by the gateway.

Because the integration was achieved largely through scripts, customisation was a simple matter of selecting or changing statements in a text file. For example, part of the purchase.cgi script defines the default currency in which transactions are to be conducted

```
# Local currency
$localCurrency="GBP";
```

This can readily be reset to cater for US dollars or Euro by replacing the above with

```
# Local currency
$localCurrency="USD";
```

or

```
# Local currency
$localCurrency="EUR"
```

In much the same way, the product description, format of reply pages and layout of order form can all be changed by the merchant. The interaction with the payment gateway, however, is fixed—it is a secure service that needs to be carried out in a specific way, as explained in Chapters 4 and 5.

The use of merchant-configurable scripts was driven by (merchant) demands for flexibility (Notting Hill Publishing) and (supplier) desire to build componentised eBusiness solutions (Norris, Davis and Pengelly 1999). The key point about this case study is that it shows that some aspects of back-end integration can be straightforward. The fact that many merchants have to use the same back-end systems makes it worthwhile building a gateway which can be accessed and used with a tactical package.

6.7 INTEGRATION STRATEGIES

Not all back-end systems will be shared by a wide variety of users. More often than not, they will be organisation-specific, and the integration of the back-end with eBusiness applications will be a task that the owning organisation has to undertake.

We are now in the realms of mainstream integration. This is a subject that has received considerable attention (albeit not enough to cure the headache that it causes). For some years, the desire to make back-end systems (sometimes called Operational Support Systems, or OSS) more

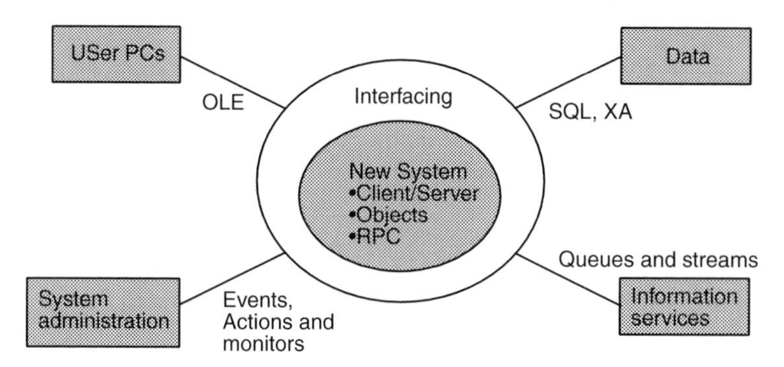

Figure 6.5 A general template for systems integration

flexible has been driven by globalisation, organisational flux and supply chain integration. In the face of this, strategies for opening back-end systems so that they can more readily be integrated have been devised.

Figure 6.5 is a general view of the various types of back-end system that might need to be integrated with a new development. The point to be made here is that whatever strategy is adopted for interfacing, there are four main types of legacy function that need to be considered:

- User PCs, where each person tends to look after their own work and attend to their own security. In terms of technology, this group use Active-X, Object Linking and Embedding, ODCB and the like to share information.

- Systems administration, where specialist units are required to look after service and network management. Bodies such as the Telecommunications Management Forum and the Internet Engineering Task Force guide the technology that is used here.

- Information services, where a huge amount of data is collected in warehouses, and these are managed to service routine reporting requirements (e.g. monthly sales returns).

- Data, which is similar to the above, with the exception that requests are *ad hoc* and need to be serviced quickly. Direct database queries (e.g. via SQL) and transaction processing are relevant here.

Mainstream computer science has put forward a bewildering array of candidate solutions for each of the above cases. Objects, the three-tier architecture, client/server, Distributed Network Architecture, Enterprise Java Beans and the Distributed Computing Environment (DCE) have all been touted as your flexible computing friend. The truth is, though, that as often as not, they can create as many problems as they solve.

In practice it is common to find captive data, trapped in a new system, that cannot readily be used elsewhere. Information systems, based on networks of computers, are littered with screen scraping fixes, tactical file transfers and process work-rounds that cost a small fortune to live with (and would probably cost a larger fortune to fix properly and realign with business needs).

So, having drawn a general picture that focuses on the issue in integrating back-end systems, we now take an end-to-end view. In the next section, we turn our attention to the front end and look at some of the approaches here.

6.8 THE FRONT END

There are a number of different approaches to building the distributed systems used for eBusiness. Each has strengths and weaknesses, and we will examine what these are for three typical cases. Our first example is of a forms-based, interfaced to a mid-tier application.

Figure 6.6 illustrates a fairly commonplace two-tier design in which the client's front-end link to the mid-tier application is provided by a browser client rather than one of the usual repertoire of clients (such as a VisualBasic client or an Oracle Forms client).

Pros and cons

There are a number of good reasons for building this style of client:

- The desktop environment is completely general purpose—just a standard browser with TCP/IP connections—there is no specific application functionality on the desktop.
- Configuration management of application software is purely a server concern, because the application has no specialised desktop components.
- A single user can make use of multiple applications from a consistent client.
- The design allows for Internet as well as Intranet access, i.e. appropriate services can be offered direct to customers.
- High levels of throughput should be achievable between the client and server. HTTP does not maintain a permanent connection to each client (even a medium-sized HTTP server can service millions of requests per day, from tens of thousands of clients).
- Rapid delivery—not a complex design.

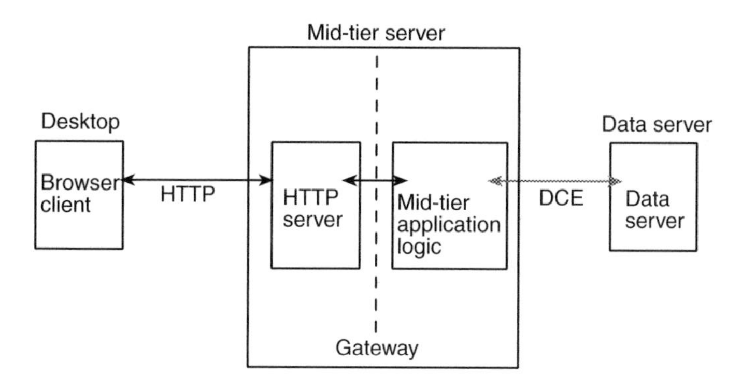

Figure 6.6 A basic two-tier design

There are, however, some drawbacks and limitations:

- The user interface at the client end is very basic: a simple forms structure.
- Although the expected throughput should be good, the latency in any individual client transaction can be poor. People acquainted with the use of the Internet are well aware that, although WWW servers may we able to service thousands of requests, the delay experienced in any particular request may be very large. Delays that are acceptable in an Internet environment may be quite unacceptable in an Intranet environment.
- There is no persistent state maintained in the client.

A simple two-tier example of this style of application can be found in many desktop directory implementations. More complex three-tier versions can be produced that allow an external customer to view and change their selection options. The mid-tier logic is usually provided by Oracle stored procedures.

A second option for eBusiness applications would be to use client end applet components to provide presentation facilities (Figure 6.7). In this case, the applets only perform presentation logic functions at the client end. Examples are:

- Data validation. Checking data as it is entered to ensure, for example, that mandatory fields have been completed and that the right types of data has been used.
- Data visualisation, e.g. tools that allow the data to be presented in a variety of visual forms such as charts.
- Data analysis. Applets may analyse and manipulate the data.

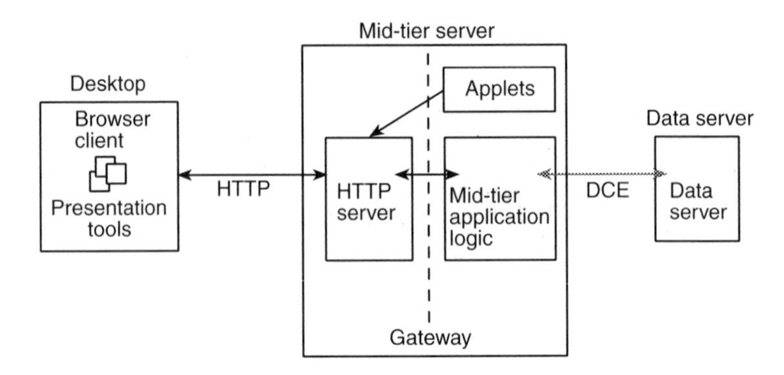

Figure 6.7 A more sophisticated option

Pros and cons

These are similar to those expressed for the first option, with the following exceptions:
On the plus side:

- A somewhat richer user interface can be offered to the user, including a number of data visualisation options such as charts and graphs.

On the minus side:

- The loading of application functionality over the network will adversely affect the performance of the system—and in particular, is likely to have an adverse effect on the latency.
- Differences in versions of browsers can lead to incompatibilities in the execution of applets. This is particularly the case where the application is considered for Internet as well as Intranet use.

In our third option, applets are not just used for presentation—they provide full application functionality at the client end. As suggested by Figure 6.8, if this is the case, there is a requirement for a more robust protocol between the desktop and the mid-tier, so that the applets can communicate reliably with mid-tier components. This would require either DCE or CORBA or some reliable message-passing protocol.

We can only scratch the surface in terms of the technical options, and it is probably already clear that there is something to suit every taste. In the authors' experience, there are few right and wrong answers, and the approach taken should be determined as much by prevailing culture as anything else. Appendix 2 provides some very practical and powerful, if

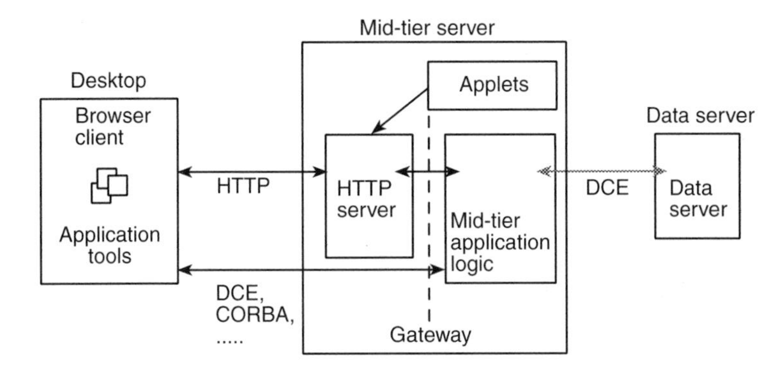

Fig 6.8 Applets provide applications

somewhat whimsical, guidelines on this very key aspect of getting the right technical approach to suit your eBusiness solution.

6.9 SUMMARY

In any walk of life, change is inevitable and computer systems probably change more than most. So, in any computer-based system, there is always going to be a mix of the old and the new, just like the contents of most people's wardrobe. For all of its gloss and novelty, eBusiness is not immune. In fact, as online trade develops, it is likely that more and more information held in legacy systems will be needed, and this means that back-end integration cannot be ignored if eBusiness is to mature and grow.

The consequence of continual change is that you either have to regularly replace everything (which is usually prohibitively expensive) or blend the new items in with the old (which is tricky but manageable). The only trouble is that the style challenge of matching clothes becomes a lot more complex when the components that make up complex, networked computer systems are the object of attention.

One of the main messages in this chapter has been that integration can be a complex and, sometimes, messy exercise for which there are precious few prescriptions. That said, we have presented some useful guidelines and strategies, borne of experience, that can minimise the challenge. In support of this, the larger part of the chapter has presented case studies that illustrate issues that emerge in the real world and hints and tips that can inform the successful integration of back-end systems into an overall eBusiness solution.

REFERENCES

Norris M, Davis R & Pengelly A, *Component Based Systems Engineering—interfaces and integration*, Artech House, 1999.

Notting Hill Publishing (http://www.dancerdna.com/)

FURTHER READING

Advanced Network Systems Architecture Handbook, ANSA consortium, Cambridge (http://www.ansa.co.uk).

Jackson M, *Software Requirements and Specification* , Addison-Wesley Longman, 1995.

Rakitin S, *Software Verification and Validation: A practitioners guide*, Artech House, 1997.

Ward J, Griffiths P & Whitmore P, *Strategic Systems Planning*, Wiley, 1993.

7

Supply Chain

Organised crime in America takes in over $40M a year—but spends very little on office supplies

Woody Allen

If you were to ask people what eBusiness was all about, most of them would cite one of the online stores that populate the Internet. Yet much of the real innovation, and most of the predicted future growth, is in the chain that leads back to suppliers of the stores and, in turn, to their suppliers.

The book seller Amazon is seen by many as an innovator in terms of their advertising, but probably their most innovative feature is the way that they handle their supply chain, behind the scenes. They have built their business by cutting out several of the conventional stages in book supply, and have merged the traditionally separate roles of retailer and wholesaler.

Some online merchants go even further than Amazon. They don't hold their own stock at all—they simply broker the supply of goods from maker to consumer. The simple motivation in doing this is to cut costs and to get a product from supplier to consumer as quickly and directly as possible. When the standard overheads are removed, there is a healthy margin left for the broker.

In this chapter we examine the logical end point of eBusiness—where one computer interacts with another to arrange for goods and services to be delivered. We start by examining one of the early attempts at supply chain automation, EDI, before going on to examine the principles, practicalities and technology that combine to enable the bright future predicted for this aspect of on-line trade.

7.1 EDI: HOW IT WORKS, WHY IT ISN'T ENOUGH

For all the hype, there is little new about eBusiness. Electronic business technology, in the form of Electronic Data Interchange (EDI), has been around for more than a decade. It addresses a fundamental and enduring business problem—the exchange of data between the computer systems that support the business operations in so many organisations. It endures as the cornerstone trading technology for of a number of successful global companies—the worldwide EDI market is currently valued at $2bn, and is set to rise to $7.5bn over the next five years.

EDI can be defined as the exchange of structured business data between the computer systems of trading partners in an agreed standard format that allows automatic processing, with no manual intervention. It is relevant to any business that regularly exchanges information such as client or company records, but is especially relevant if you send and receive orders, invoices, statements and payments. In terms of the full spectrum of eBusiness, EDI fits neatly in the supply chain area, as illustrated in Figure 7.1.

EDI allows transactions that have required paper-based systems for processing, storage and postage to be replaced and handled electronically—faster and with less room for error. This is why it is sometimes called 'paperless trading'. EDI can be thought of as analogous to email, except that the transactions take place directly between computer systems instead of human beings and, because of this, the information needs to be more rigorously structured. As with email, some form of network is placed

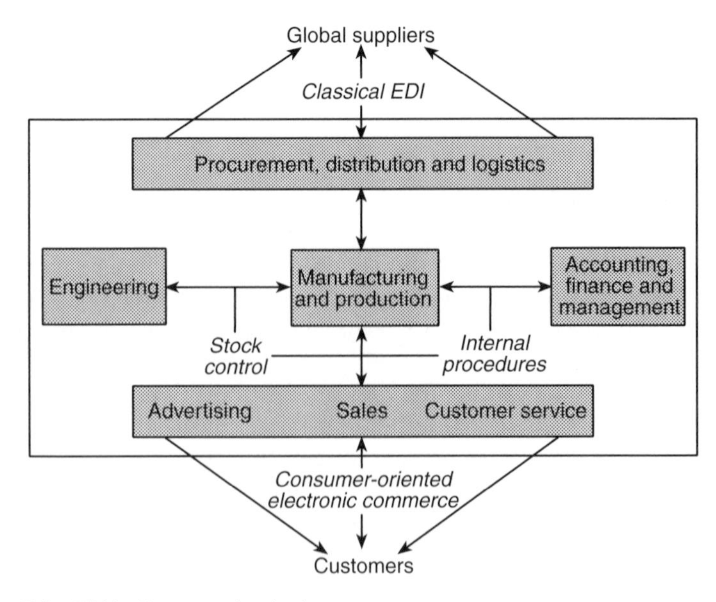

Figure 7.1 EDI in the supply chain

between two (or more) trading partners to store and collect messages, provide an audit trail and resolve communication protocol differences. These networks, known as Value Added Networks, tend to be proprietary (such as EDI*Net on the UK), rather than open (like the Internet). They provide services that specifically support the carriage of EDI interchanges.

These EDI interchanges may contain many different document or message types. The structures of the interchanges and message types are defined both by national and international standards. In addition, some industry sectors have created their own standards—ODETTE, for example, is used in the car construction industry, and TRADACOM pervades the UK retail business.

The industry-specific standards have meant that you could only communicate electronically with other companies using the same EDI systems and processes as yourself. Given this limitation, international standards such as ANSI X.12 (in the US) and the global UN/EDIFACT (published as an International Standard, ISO 9735) have been advanced to ensure interworking. The latter has published over 200 message types, constructed from a common set of 'building blocks' or segments.

All of the EDI standards are based around five interrelated areas: Messages, Segments, Data Elements, Syntax, and Message Design Guidelines. Any EDI message, such as an invoice, is compiled from standardised interchangeable elements called segments, which are composed of data elements. Segments are related to specific objects or entities within the message and can be thought of as the fields of a database record. Examples of a segment would be a company name, an address, a delivery location and a specific product. Segments consist of strings of data elements. These elements are the basic vocabulary of EDI, and they constitute the smallest item of information in an EDI message. A typical data element would be a postal or zip code, the quantity of items ordered and product codes.

EDI syntax defines the grammatically correct ways of combining the data elements and segments into a message. For EDI transactions to run efficiently, codes are used to identify products and locations, as these are more accurate and less prone to misinterpretation than textual descriptions. Message design guidelines give directions on how to build messages to answer the particular business needs of any given industry.

Because of the diversity of business that EDI can support, industry segments have established special-interest EDI groups to ensure that the standards meet their needs. The ANSI X12 standards, for instance, contain several subsets such as the Uniform Communication Standard (UCS), used primarily by the grocery segment, the Voluntary Industry Communication Standard (VICS), used primarily by general merchandise retailers, and the Warehouse Industry Network Standard (WINS), used by warehouses. There are a host of others to cover virtually everything from telecommunications to books and defence.

To illustrate what sort of detail is contained in these subsets, we can look at X12F, which deals with banking and finance. Inside X12F, there are 37 standard messages that have been released for use, and another six that are still in development.

Some of the most important of these:

- 820-Payment Order/Remittance Advice, which establishes the data contents of a payment order or remittance advice transaction set. This can be used for three different purposes;
  1. to order a financial institution to make payment to payees on behalf of the sending party,
  2. to report the completion of a payment to payee(s) by a financial institution,
  3. to give advice to the payee by the payor on the application of a payment made with the payment order or by some other means.
- 821—Financial Information Reporting, which is used to transmit a detailed balance, any service charges, and any adjustment details from a bank to its corporate clients or between or within corporations.
- 827—Financial Return Notice, which is used to report to the originator the inability of the originating financial institution to process a Payment Order/Remittance Advice Transaction Set (820).
- 828—Debit Authorisation, which provides authorisation to withdraw funds from an existing account.
- 829—Payment Authorisation Request, which provide the authorisation to withdraw funds from an existing account for the purpose of paying an invoice or vendor.

For each of the above, there are segment and data element definitions, along with prescribed syntax. With X12 covering such a wide range of applications, its full breadth is quite impressive.

The other major international EDI standard is EDIFACT, which has been developed by the United Nations and is published by the International Standards body, ISO. EDIFACT provides a globally uniform set of rules for formatting business transactions, so is well suited to multi-industry and multi-country computer data interchange. Compared with all other EDI formats, EDIFACT has two important advantages. The first is the global validity of the EDIFACT standard, and the second is the very extensive multi-industry range of transactions for which message types are available. These message types are developed for all transactions occurring in companies from different industries, for instance orders, delivery status messages, invoices, payment orders, or customs formalities.

The operation of a large organisation such as British Telecom gives some idea of the scale of commercial EDI implementations. Twenty different parts of BT use EDI in their day-to-day business, the largest generating around 10 million transactions per annum. The introduction of EDI has led to massive stock reductions and was a key enabler for a 'single warehouse' strategy. The BT implementation also supports their 'Direct Fulfilment' service, the channel whereby customer equipment orders are guaranteed next day delivery anywhere in the UK if placed before 5 pm.

The invoice and payment cycle associated with a large proportion of these orders is also automated with invoices received electronically via EDI and automatically matched against the original purchase order. If there are no discrepancies and receipt of the goods has already been confirmed, payment is automatically made via the Banks Automated Clearing System (BACS). Thus, beyond the initial order entry, there is no manual processing at all.

All of the interest and activity in eBusiness has breathed new life into the EDI community, and a number of new industry groups, each with its own standards for electronic commerce, have emerged (in the US alone there are as many as 24). RosettaNet is a good example of the genre. Whilst recognising that EDI and bar code standards present a foundation for electronic transactions across many industries, this body is suggesting that it is time to complement them with industry-specific business interfaces that not only address basic transactional data, but encompass all aspects of commercial relationships. In much the same vein, the Open Buying on the Internet (OBI) consortium are driven by a shared vision to facilitate the rapid implementation of Internet-based electronic commerce. The first version of the OBI standard was released in 1998.

All of these new standards draw heavily on the established principles for common business processes in EDIFACT and ANSI X12. The technical basis being proposed is the Extensible Mark-up Language (XML), which does not define particular business document formats, but is a meta-language that defines the syntax for a particular document type. With XML, a Document Type Definition (DTD) is built from a particular set of 'tags' and 'tag attributes' that identify the nature of the content. (See the end of the chapter for a blow by blow account of XML.)

It is salutary to reflect that, despite 15 years of government and industry advocacy, EDI has attracted only around 100,000 commercial customers worldwide. The key stumbling blocks have been cost and infrastructure and, specifically, the fact that most EDI users must connect over dedicated, proprietary networks or VANs. The Internet removes this impediment and allows much of the thought that has gone in to EDI to find its home.

7.2 WORKFLOW

If you are to have any hope of automating a supply chain, the links in that chain must be well understood. So it is important to know precisely what the steps in a given transaction are, who is responsible, and what needs to be done to move from one step to the next.

Workflow addresses this. It is all about stringing together the processes in different organisations and creating a clear end-to-end path. In many cases, this is little more than systematic commonsense. For example, the seven (yes, seven!) steps in ordering a box of paper in a large organisation would be:

1. Gets OK by boss (verbal).
2. Then by finance person (written).
3. Send finance authorisation to buyer (who owns relationship with company and will handle it within the context of a call-off contract).
4. Order goes to supplier.
5. Goods come to originator, who fills in a Goods Receive Notice (GRN).
6. GRN is sent to accounts payable, who will also have a copy of the original order. They pay when they get an invoice from the supplier.
7. They make Bank Automated Clearing System (BACS) payment to the supplier and the real money flows.

This approach can be used for a variety of processes, and (because many real-life process are quite complex) the results are usually presented graphically (Figure 7.2).

Whatever form the workflow description takes, the vital information being captured is what order things are done in, what data has to be created or transmitted, and what processing happens at each step. The challenge is to produce an electronic system that is functionally equivalent but more efficient in terms of

- speed to complete
- resilient to fraud
- auditable
- less wasteful (e.g. uses little or no paper).

There are commercial products such as Oracle Workflow, SAP, Lotus Notes and many others that can readily automate a defined workflow. Indeed, there is more than enough technology to choose from. The difficult bit is getting the workflow definition right, i.e. so that real people can work with it and that it satisfies the four criteria stated above.

Some of the key issues in workflow are:

- *Application of electronic signing*. With paper orders, we would sign by hand at every stage (as any Fedex user knows). Many systems replicate this by username/password protection, but this is not a strong enough

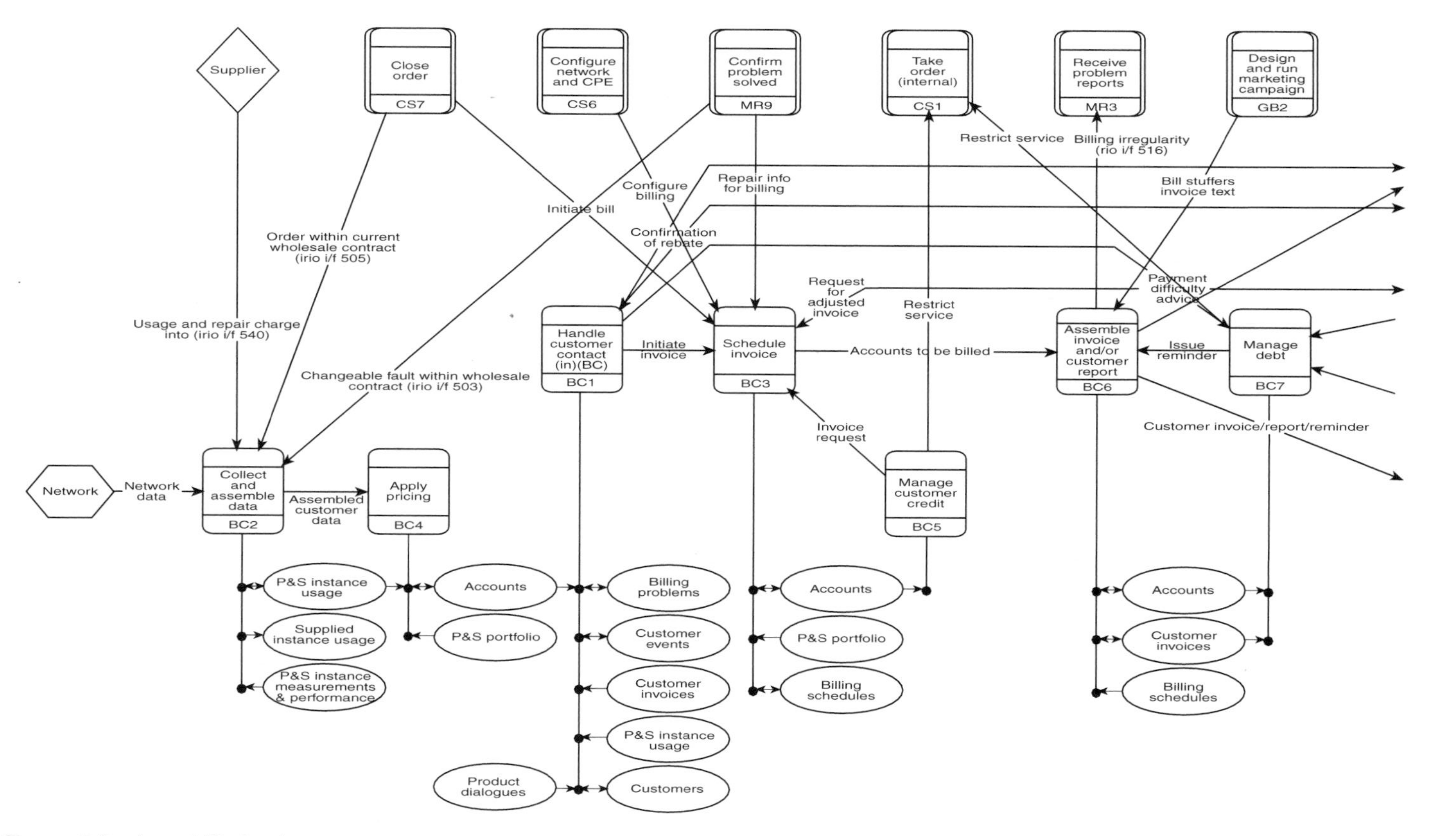

Figure 7.2 A real-life business process

mechanism (it is rather like signing for all goods at the front gate of a factory, irrespective of which department they are for). The expectation is that the medium to long term solution will be the use of personal digital signatures (probably stored on smart cards or provided through USB port dongle).

You made need different signatures for different roles. In the UK doctors have two signatures—one for prescription and a different one for personal transactions. The same would be true for digital signatures. For instance, a finance director has different authorisation levels depending on whether it is a capital purchase or entertainment expenses that are being authorised. It is quite common for the signing limits to be millions for other peoples' kit, but next to nothing for personal use!

- *Design of workflow.* This is much like the design of any distributed system. To that end, it is important to realise that the same aims, principles and practices apply. For instance, a good workflow design should work irrespective of the location of its components (this is transparency of locations, an aim for designers of distributed applications.

  An essential principle in workflow is that there should be no doubt about where you are in the workflow, and it should be possible to backtrack if things go amiss. To maintain the integrity of the end-to-end flow, no part should be indeterminate. Four properties, atomicity, consistency, isolation and durability, should be maintained for all transactions. The first two ensure that a transaction completes or it fails (there is no room for 'not sure'), the third that one workflow doesn't impact or get affected by another, and the last that the workflow only moves forward (there is no backtracking).

- *Standards.* Workflow doesn't stop at the boundary of any one company or country—it relies on others playing the same game to the same rules. Hence, the invoice sent by the supplier needs to be handled alongside the purchase orders that you have generated, so they must conform to certain standards of content, delivery method and so on. More on this in Section 7.11.

  In terms of automating a workflow, there is no reason to do anything other than buy a ready-made solution. The accepted next step is that you move your business process to match (rather than 'customise' the purchased technology in any way). So, if you have a company policy that seven people have to sign off a major purchase and the system only supports five, it is probably cheaper and easier to change the process than to customise the systems (note—this is a real example, and one that reflect the accumulated wisdom across the computing business).

  There is a significant implication here. The adoption of Commerce-1 or SAP/R3 (or of any form of supply chain automation for that matter) in any large, established company is that a big commitment has to be made in moving to this way of trading. It is more than just a new suite of software that is being introduced—existing processes are being abandoned (perhaps with no return), and the whole company is moving into an unknown way of operating. Some of the most respected voices in the IT business have stated that eBusiness is a major and irrevocable change to the *status quo*. The consequences of workflow automation illustrate their point.

7.3 DIRECT AND INDIRECT GOODS

One of the first things to do when considering supply chain automation is to clearly distinguish direct and indirect goods. The former are those that you need because of the business you are in, the latter those that you need because you are in business. So office PCs, stationery, company cars, furniture, buildings, etc. are indirect and direct goods are specific to the industry—ball bearings, etc.

When automating the supply chain, there are differences in the way you handle these two sorts of procurement. Indirect orders are generally initiated by a non-specialist, but are within the context of a contract negotiated by a professional buyer. The direct goods are obtained by someone who is an expert in the goods being supplied; they understand the nature of the product being acquired. The implication on the systems is that indirect goods are loosely specified and price is what people look at (e.g. biros have flat tops, rounded tops, fluting, etc., but the purchaser only really cares about the colour and cost).

When most people buy paper, they only want be sure that it is A4 and in packets of 100. A professional paper buyer wants to know about weight, finish, exact colour/shade, acid content, recycle content, porosity. Just the same as a professional cable buyer, laser buyer or fine art dealer want to know all sorts of technical detail. The point is that selling a product depends upon who you sell to.

In any complex business, the telecommunications and computer industries, for instance, there is most certainly a mix of supply approaches that need to be considered. Looking at what many of the leading players in this area do, it is generally the case that partnerships are forged to cope with complex, high value, bespoke and high risk ventures. In these instances, the customer cannot readily play the market for the required components and suppliers cannot deliver the goods via an established production line. So mutual dependency is high as is the cost of change for both parties (resourcing a key component for the customer, finding a new purchaser of specialised goods for the supplier).

For more mundane, commodity items, most opt to purchase from the cheapest or most convenient supplier. The relationship is based on simple transactions, and the cost of change (for the customer) is low. In this instance, the challenge for the supplier is to compete on cost—product requirements are known and one supplier's offerings are much like any others.

Selecting the right option for a particular project is not that straightforward. It is quite viable to establish a partnership for commodity items or to play the market for a leading edge system. Both approaches have their pros and cons; the trick is to maximise the pros and minimise the cons.

How the two main strategies that can be applied is well illustrated in the car industry, where companies like Toyota and Volkswagen have (deliberately) taken very different positions. The former have opted for long-term partnership, the latter for short-term optimisation. This is illustrated in Figure 7.3.

At the two extremes are Toyota and Honda, who commit to a group of suppliers in return for ongoing price, quality and time improvements in their performance, and Volkswagen and General Motors, who are happy to ditch a supplier if a cheaper or more convenient alternative comes along. It should be noted that the partnership option is by no means based on implicit trust. The aim, in all cases, is to reduce overall cost and, in practice, benchmarking is still carried out and prices are in line with the wider market.

Between the poles have (at times) come Ford and Rover. The former has sometimes drifted with no fixed long-term strategy, the latter has been in transition. Although difficult to prove, the general perception is that those whose strategy and behaviour match (i.e. those at the poles) are getting a better service from their supplier base than those drifting or in transition.

This picture may tell us something about the market, but there are some cautionary notes. The main one is that the car industry has a few big customers and many small suppliers. The dynamics of the information technology industry, for instance, is a bit different—suppliers and customers are of similar size and there are few players overall.

Notwithstanding, there are areas where a specific supply strategy should be pursued, and Figure 7.4 tries to illustrate this. If we move away from the polarised view to a more discriminatory one, there is a segment of most businesses that benefits from collaboration.

The axes on Figure 7.4 are *co-ordination* (a measure of the required control that needs to be exercised) and *incentive* (a measure of cost sensitivity).

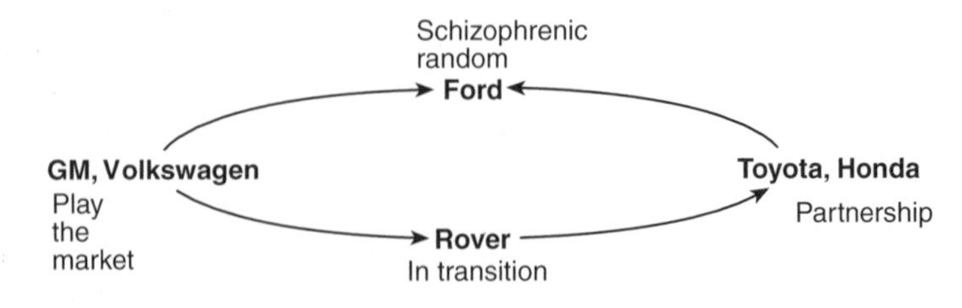

Figure 7.3 The polar approach

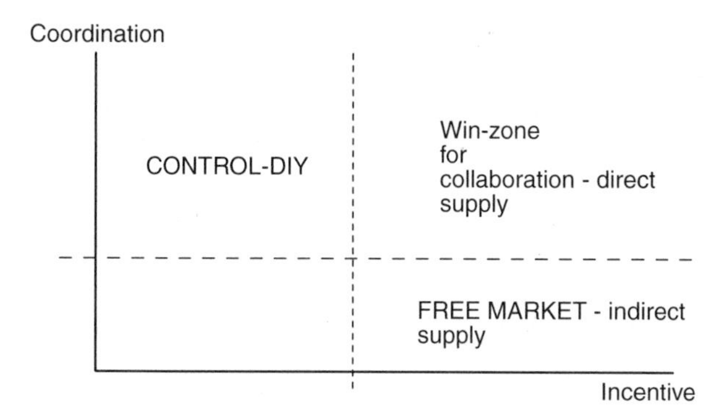

Figure 7.4 The collaboration win-zone

When price is paramount, the free market applies and the cheapest supplier wins. This is clearly the province of indirect supply. When there is some flexibility on cost and some willingness to relinquish direct control (i.e. trust), then collaboration comes into its own, and we are in the province of direct supply.

Raw materials for manufacture that you might think of as commodity or indirect are increasingly being differentiated by their packaging and presentation. This is because people now look for goods that are specially packaged so that they can readily be used in their environment with minimal handling. For example, many of the high street retailers have certain shelf sizes, so look for packed goods that fit directly onto them.

The electronic systems have to recognise these different purchasing options and subtleties so that they enhance and streamline the process. The focus so far has been on indirect goods as this is an easier target for automation—catering for every aspect that an expert purchaser might be interested in is both difficult to do and, due to the specialist market, difficult to get a good payback from!

One little terminological point before we leave this topic is that indirect goods, as described here, are referred to as MRO (Maintenance, Repair and Operations) in the US. In truth, they are one and the same thing.

7.4 NON-CREDIT CARD PAYMENT SYSTEMS

Credit cards are so much a part of modern life, why would we be thinking about trading without them? The simple reason is that the type of trade we are concerned with here is not personal—it is company to company. Also, credit cards are optimised for the one-off transaction. In the supply chain case, the two parties have a long-term relationship in which bulk discounts, call-off contracts and preferred supplier status play a major part. One further issue that we need to consider is that the value of a transaction is often not clear with Invitations to Tender and call-off contracts. The cost of a credit card transaction is high—what we want is a cheap, volume means of charging, invoicing and settlement—and one that can cope with a measure of uncertainty.

We are helped here as there is a high level of trust. The safeguards and checks undertaken with a credit card are not all necessary in this instance. The terms are net rather than transaction specific, and there is often a sliding scale of charges dependent on volume. Rolling settlements are often required to support a long-term trading relationship. An even simpler mechanism is the standing order against expected spend (e.g. management consultants tend to be used by large companies on a regular basis. Rather than negotiate every job, it is easier to make a quarterly payment based on past volumes, and carry forward the difference between forecast and actual).

Whatever the mechanism, the professional buyer role needs to be supported. This person is not necessarily the one who makes the purchase, but they do negotiate the bulk deal. Much of the trading relationship will be set up outside the electronic realm, so the payment system must accommodate individual purchases in the framework of this off-line framework.

Once all of the necessary internal adjustments and reconciliation have been made, the balance is paid either as a BACS transfer or by printing a cheque. In terms of process, this is only a small part of the overall picture (even if it is a single £1M cheque that covers 100,000 individual orders). These Mega- and Tera-payment systems (cf. the micro- and nano-payment systems we saw earlier) are all about aggregation of payments to minimise overall time and effort. In practice, that aggregation might be done on a spreadsheet or the back of a cigarette packet. The important issue in all this is to make sure that the overall tally is accurate—the paying company needs to have auditing set up to show that the cash that has gone out matches the volume of goods in (at bulk level), as does the receiving company.

7.5 MULTIPLE SUPPLIERS

Often, one buyer will have many suppliers—a multinational can have thousands and a national supermarket chain will probably have several hundred. These buyers are in danger of information overload because they need to view a lot of catalogues, one for each of their suppliers, before choosing what they want. The real complication is that several of these suppliers will carry the same line (ball point pens, for instance) which the buyer needs to compare for cost, quality and availability. The trouble comes when catalogue entries for essentially the same thing—the humble ball point pen—appear in the guise of a pen, a ballpoint, a biro and a rolling writer! It can be hell!

When it comes to indirect goods (or MRO), the same issue is being addressed by numerous buying organisations, and there are two views on how best to tackle the problem. The first is to use a market site that mediates between all of the players. Companies like Commerce One provide a managed common area to which buyers and suppliers alike can connect and do business. The second view is that all suppliers should run conformant systems, so there is no need for mediation. Data would be kept in an Enterprise Resource Management System such as SAP/R3, and could be used by anyone conforming to that standard.

In essence, these are (respectively) the heterogeneous and homogeneous approaches. Either trust in standards or insist on conformance. The aim is the same in both cases—given that suppliers offer their goods through some form of catalogue, the buying systems must be able to view the aggregate of these catalogues because they will want to purchase by category of goods rather than suppliers. So, if a box of pens is what they want and they are not bothered whether it is from one supplier or another, they simply want to see a comparison of prices, delivery dates and stock availability.

Taking the first of our candidate solutions, some service providers operate market sites that pull in data from all of the separate suppliers and 'normalise' it so that they offer a single virtual catalogue to buyers. This illustrated in Figure 7.5. In the figure, the addition of a new buyer to the established community is straightforward, rather like adding a new user to a Local Area Network—you only have to plug them in and register their name and details on the site management system. Because this solution builds many-to-many connections, it not only solves the multi-supplier problem, it also (at a stroke) solves it for the whole community of buying organisations—if it is a good enough solution for Steve-Mart's indirect goods, it is probably good enough for Phones-R-Us as well.

The main role of the market site is to build an intermediate catalogue

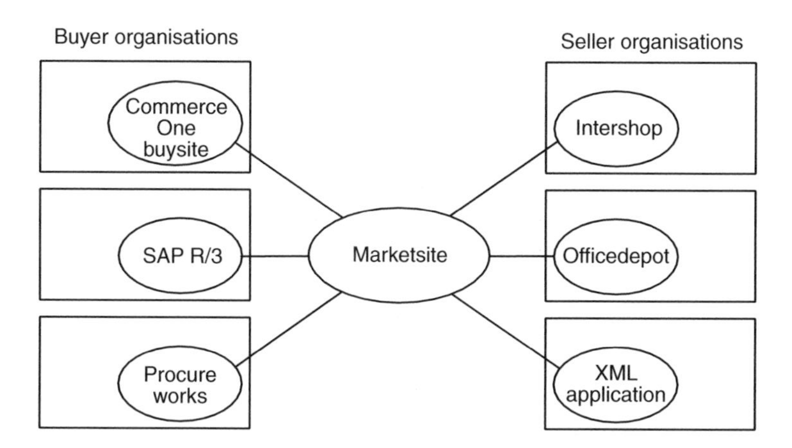

Figure 7.5 The structure of a MarketSite

that can be used by all participants. This has to be able to take data in as a variety of different forms—comma or tab separated flat files, SQL, various flavours of XML, etc. This is a non-trivial challenge, and one that gives considerable impetus for a single solution, rather than having each buyer building their own. Hence, the market site adds considerable value from the buyer's point of view.

Our second option is that all the vendor catalogues are standardised in a common machine readable way, and are presented through an industry standard interface. This would allow the buyer site system to pull data from all of its supplier's systems and present it to the purchaser in a consistent way. Because this option entails both buyer and supplier adopting the same standards, it puts more onus on both parties—the light at the end of the tunnel may be bright, but the work in getting there is significant and cannot be put onto a third party, as it can with a market site solution.

The reality is that the first option is the pragmatic solution, and the second is perhaps an ideal, but one that is not yet with us (and may never be). Indeed, we already see a situation in the market where different component suppliers are proffering different standards. Experience in the computing world suggests that it will get worse before it gets better!

Also, there is the assumption that there is some consistent structure that makes it possible to have a machine readable format that can pick out all of the 4 mm blue biros, 60 Watt soft-glow, bayonet fitting light bulbs, etc. In truth, this is a non-trivial job, so the intermediate site (which is, of itself, a fairly complex entity, built and updated from a variety of suppliers) may well be the state-of-the-art for some time yet.

7.6 CASE STUDY—Commerce One

Commerce One are a West-Coast US company who specialise in supply chain automation for indirect goods (known as Maintenance, Repair and Operations, MRO). Their starting point is that Enterprise Resource Management (ERM) applications from SAP R/3, Baan and J D Edwards have taken much of the costs and inefficiencies out of the internal workings of companies and so now the next prize is to do the same for the supply chain.

The Commerce One approach is to develop a many-to-many (suppliers to purchasers) trading model with three components:

- the 'Buysite' operated within a corporation that buys goods through the system;
- the 'Seller Site' operated by selling organisations such as a stationery supplier, office equipment companies, and so on;
- the 'MarketSite' which mediates between buyers and sellers.

Their vision is that there should, globally, be just a small number of market sites (corresponding to particular geographical areas or, perhaps, some specialist buying communities) which provide the open markets for MRO goods. The intention is that the market sites employ multiple standards to allow many different buyers and sellers to undertake transactions through the market.

The operations offered by their MarketSite product fall into two main areas: those concerned with catalogues and product data; and those concerned with transactions for the goods ordering process.

In terms of the catalogue and the management of product data, on the supplier side, the MarketSite product can pull in catalogue data from a range of industry-standard catalogue products. These include Intershop, Microsoft Commerce Server, Office Depot and various XML-based applications. Also, it can import data in less esoteric forms such comma-separated text files. This data from multiple suppliers is aggregated so that it can appear to buyers as a single virtual catalogue. Each buyer organisation will, however, have a customised view onto the catalogue so that they only see the products of the subset of suppliers with whom they do business. Also, typically, some details such as prices will differ according to which customer organisation is viewing the MarketSite. On the buyer side of the MarketSite, again an open interface allows a range of procurement applications to access the site. These include Commerce One's own BuySite application, ProcureWorks from RightWorks, SAP R/3 and others.

As well as perusing MarketSite's virtual catalogue, the buying organisation needs to be able to conduct transactions for the ordering of goods. These transactions may be handled by the same applications (BuySite, SAP, etc.), but may, on either the buying or selling side, entail other technologies such as EDI messages, SMTP electronic mail, OBI (Open Buying on the Internet) messages, or plain vanilla fax. Marketsite aims to offer a single point of interoperability for the interchange of all these message types.

In addition to its two main interfaces, those to buyers and sellers, MarketSite offers a range of other commerce-related interfaces:

- Shipping and logistics services such as UPS, allowing, for example, purchasers to track the progress of their goods.
- Payments gateways. These are particularly to credit card companies which are quite widely used, even for procurement by large corporate organisations in the US, but are probably less appropriate to European buyers.
- Taxation services, to apply appropriate local tax rates to selected goods.

Commercially, Commerce One have been successful in establishing relationships with telecommunications providers to host market sites in key world markets. For instance, Commerce One are allied with Worldcom in the US, BT in Europe and NTT in Asia. In all of these markets, a large number of high street stores, well-known brands and familiar businesses are on a Commerce One site for the supply of their indirect goods.

7.7 INTRANETS, EXTRANETS AND COINS

So far, we have assumed that the Internet provides the network infrastructure upon which eBusiness is built. It is not an exclusive option, and there are other networks (usually ones with an added degree of security or performance) that can be used. Most notable are private Intranets, and closed community Extranets and CoINs.

Perhaps the most concise way to define an Intranet would be to say that it is the deployment of Internet technology to meet the needs of particular group or organisation. In operation it satisfies the same need as a Virtual Private Network or Enterprise Network. Because of this (and despite being built on the same underlying technology as the Internet), an Intranet is quite different. The fact that it is built to the requirements of a particular set of user means that performance, security and quality of service guarantees can be designed in. The basic setup of an Intranet is shown in Figure 7.6.

In reality, there would be numerous computers, servers and Local Area Networks connected together. Typically, an Intranet constructed along the lines of that in Figure 7.6 would provide a range of information services to user terminals. News feeds, mail, file access and online references would be provided, along with a host of other information services. In many organisations, the Intranet provides the main source of working

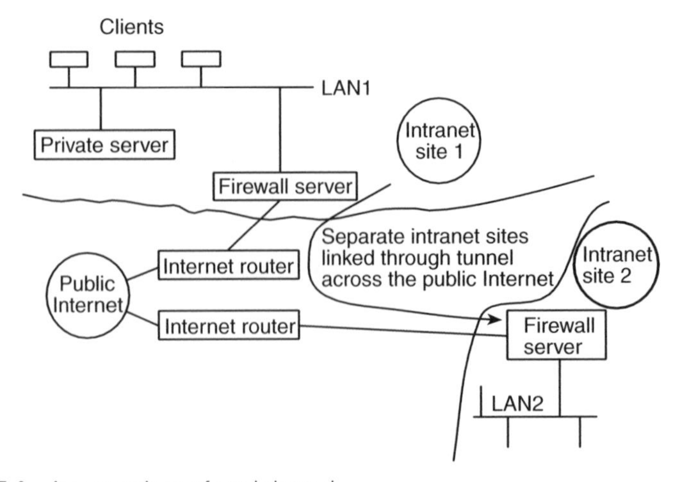

Figure 7.6 An overview of an Intranet

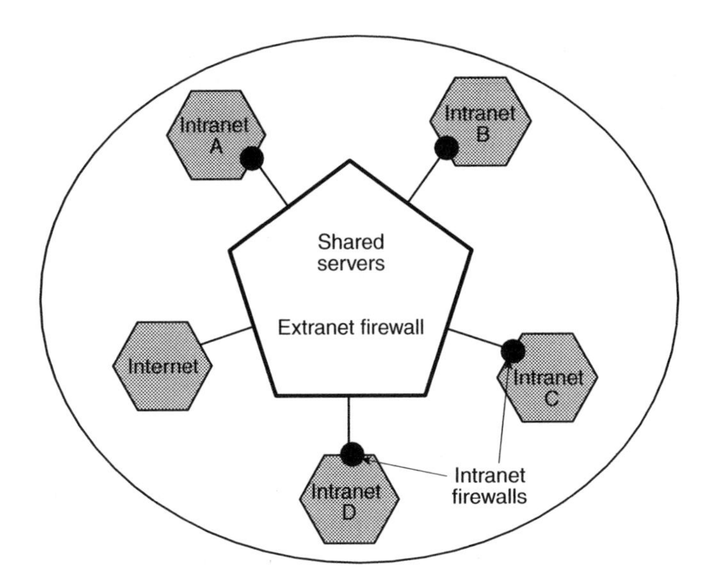

Figure 7.7 Intranets and Extranets

material, the vehicle for cooperative projects and the preferred reporting, accounting and supply route.

Intranets can be extended to form an Extranet or Community Of Interest Network (COIN) that allows other organisations to connect to shared resources (directories, files, etc.). When this is done, normal security concepts have to be extended slightly, as shown in Figure 7.7.

Instead of having a secure zone behind the firewall and an insecure public zone, there is a third area, referred to as a demilitarised zone (DMZ). In general, this calls for no more than a reconfiguration of firewalls, although in practice, all parties need to co-ordinate their policies and practices.

7.8 LEGACY SUPPLIERS

We saw in Chapter 6 that you cannot (and, indeed, should not) ignore installed equipment, channels and systems. In just the same way, there will always be those suppliers who do not fit in with the new pattern, and who are better accommodated than ignored.

The simplest way to retain links to those who cannot or will not trade online is to maintain legacy communication links. It should be possible to integrate orders via fax, etc., even if manual intervention is necessary. It is likely that some form of gateways will be needed into EDI, fax and all those other order and supply mechanisms that will be around for some years yet. The reasons for this prediction are as follows.

With the focus of supply chain systems being on the large-scale buyers, the small buyers will inevitably follow, but not all at the same speed. Significantly, and in sharp contrast with the supplier to consumer end, the cost of entry into the eBusiness supply chain area is high. Put this together with other barriers to entry (most notably, the fear of ditching tried and trusted processes, concerns over being sidelined in a more dynamic world and possible backlash from teething problems experienced by the large companies) and there will be a spectrum of legacy suppliers.

7.9 XML

A point worth restating at this point is that widespread supply chain automation depends upon high volume computer-to-computer interactions which, in turn, depend upon standard formats for the information being exchanged. The great hope in this area is XML, which provides a standard language for supply chain interactions.

Most people are familiar with the HTML standard, which allows the structure of a document to be expressed in a way that can be interpreted by a variety of web browsers. For example, the HTML language includes the tags <CENTER> and </CENTER> which delimit a section of text that should be centred on a screen. XML (eXtensible Markup Language) takes this a stage further, and permits the definition of tags for many business-related fields such as tags for dates, names, addresses, prices, and so on. Hence, it is possible to produce an electronic document or online form that can be displayed to a user and can also be read by applications which can easily pick out the key data fields.

Although XML provides the framework for defining these form-based fields, standardisation work is still needed to agree common standards for the field data. This is by no means straightforward, and the emergence of rival standards will lead to the development of separate families of e-commerce systems, only able to interoperate with other members from the same family. This could fragment the market and slow the take up of XML for business information storage and transmission.

To start at the beginning, HTML, XML and related languages are based on the Standard Graphical Markup Language (SGML), an ISO standard of 15 years or more standing which allows documents to be marked up to indicate their structure. SGML does not define tags such as the < CENTER> and </CENTER> examples above, but rather is a 'meta language' for defining these and similar tags. To create a new markup language from SGML, one must create a Document Type Definition (DTD) which uses the SGML structure to define the tags that make up the new language. For example, HTML has its own DTD.

SGML is usually perceived as excessively complex, and hence XML was conceived as a cut-down version of SGML. Nevertheless, it retains the ability to create new tags through the application of DTDs. HTML has a fixed 'tag' set to indicate basic document structure such as paragraphs, presentation such as bold, and even both at once, such as headings. XML allows the production of DTDs that are a specification for a set of tags (and attributes than can be given or implied with each tag) and the order in which they are allowed to occur (Figure 7.8).

As an example of the use of XML, let's look at a simple memo. This could be a fairly standard form used regularly by a garage in communicating with its customers. The desired result for printing or display on a screen is shown below.

13 March 99

To: Mr A Customer
From: Mr J Fangio
Subject: MOT
I am pleased to inform you that your Rover car (registration J111MRT) is ready for collection.
May I remind you that you should have your car serviced again in 6 months from now.

The next illustration shows an HTML document that will achieve this. If you are familiar with HTML, you may be able to devise a simpler way of achieving the same result, but you will note that most of the HTML constructs are used to control the presentation and layout. Actually, finding the useful data is not so easy, even for a human being. It would be a complex task, indeed, to devise a computer program that could extract data such as the customer data or the date.

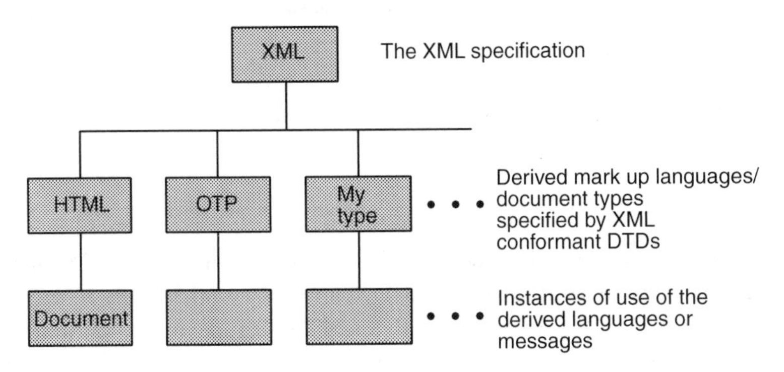

Figure 7.8 The XML language hierarchy

```
<!DOCTYPE HTML PUBLIC "-//W3C//DTD HTML 3.2//EN">
"http://www.w3.org/TR/REC-html3.2/ html3.2.dtd">
<html>
<head>
<meta http-equiv="Content-Type" content="text/html; charset=iso-8859-1">
<title>MOT memo </title>
</head>
<body text="#000000" bgcolor="#FFFFFF" link="#0000EE" vlink="#551A8B">

<TABLE WIDTH=80% CELLPADDING=2 CELLSPACING=2 BORDER=0>
<TR>
<TD WIDTH=60% VALIGN=MIDDLE> <P> </TD>
<TD WIDTH=20% VALIGN=MIDDLE> <P ALIGN=CENTER> <CENTER>13
March
99</TD>
</TR>
</TABLE>

<TABLE WIDTH=40% CELLPADDING=2 CELLSPACING=2 BORDER=0>
<TR>
<TD WIDTH=20% VALIGN=LEFT> <P> <FONT SIZE=2>To:</FONT> </TD>
<TD WIDTH=20% VALIGN=MIDDLE> <P ALIGN=CENTER> <CENTER>Mr A
Customer</TD>
</TR>
<TR>
<TD WIDTH=20% VALIGN=LEFT> <P> <FONT
SIZE=2>From:</FONT> </TD>
<TD WIDTH=20% VALIGN=MIDDLE> <P ALIGN=CENTER> <CENTER>Mr J
Fangio</TD>
</TR>
<TR>
<TD WIDTH=20% VALIGN=LEFT> <P> <FONT SIZE=2>
Subject:</FONT> </TD>
<TD WIDTH=20% VALIGN=MIDDLE> <P
ALIGN=CENTER> <CENTER> <B>MOT>/B> </TD>
</TR>
</TABLE>
<P>
I am pleased to inform you that your Rover car (registration J111MRT) is ready for
collection.<P>
May I remind you that you should have your car serviced again in 6 months from
now.<P>
<img src="/gifs/logo.gif" alt="Company Logo" height="48" width="307
</body>
</html>
```

In contrast, we show below an XML version of the same information. Even if you are not familiar with XML, it should be fairly easy to pick out the useful information. There is actually one bit of information included—the recommended date for the next service—which is not printed, but is included as useful data for the garage's use. You will notice that it can be arranged for the key pieces of information to be identified by their own tag or attribute, which makes it easier to write a program to find and process the information. For example, a computer that needs to find the date in this message need only find the <DATE> and </DATE> tags.

```
<?xml version="1.0" encoding="ISO-8859-1" standalone="yes"?>
<memo>
<to>Mr A Customer</to>
<from>Mr J Fangio</from>
<date>13 March 99</date>
<subject>MOT</subject>
<text>. I am pleased to inform you that your <car registration="J111MRT">Rover
</car> is
ready for collection.</text>
```

```
<text>May I remind you that you should have your car serviced again in 6 months
<next__service date="13/09/1999" />from now.</text>
```

```
<graphic/>
</memo>
```

The next illustration shows the Document Type Definition (DTD), which was used to constrain the generation of this message. When it is re-applied to the message it can be used to check the syntactical correctness of the message (can be useful for detecting certain errors), and the DTD can add some extra fixed information that aids subsequent processing.

```
<!DOCTOR memo [
<!-- This is a document type definition for a simple garage memo -->
<!ELEMENT memo                (to, from, date, subject?, (text|figure)+)>
<!ELEMENT text                (#PCDATA|car|next__service|citation|figref)>
<!ELEMENT citation            (#PCDATA) >
<!ELEMENT car                 (#PCDATA) >
<!ELEMENT next__service       (#PCDATA) >
<!ELEMENT figref              (#PCDATA) >
<!ELEMENT figure              (graphic, caption?) >
<!ELEMENT graphic             EMPTY) >
<!ELEMENT caption             (#PCDATA) >
<!ELEMENT to                  (#PCDATA) >
<!ELEMENT from                (#PCDATA) >
<!ELEMENT date                (#PCDATA) >
<!ELEMENT subject             (#PCDATA) >
<!ATTLIST car registration    CDATA "not available">
<!ATTLIST next__service date  CDATA >
<!—The date should be entered in the format dd/mm/ccyy -->
]>
```

An internationally recognised consortium known as the W3C (World Wide Web Consortium) is developing the syntax for XML itself, and some other 'members of the supporting cast' such as the style sheet language (XSL), a linking language (XLL), a schema language and a Document Object Model. The standards are all relatively new and immature—indeed, some have yet to properly emerge, and there are more to come—but there is international agreement on these base standards. There the good news ends, for beyond that it is chaos!

Two different groups have produced a set of XML DTDs for enabling eBusiness interactions. Microsoft are producing a set for their Office suite of applications. Other groups are setting up industry specific activities. For instance, in the UK the Travel Technology Institute is considering establishing a set of data definitions and DTDs for the travel industry, and CEN/ISSS have workshops doing the same for the sanitary ware industry and the health care procurement industry. These industry specific efforts will be fine so long as there is only one set of agreements per industry, and they make use of a common set of basic data definitions and generic messages (such as a generic Order message).

What is conspicuous by its complete absence at present is any recognised international effort to agree a common set of basic data definitions and generic messages, such as a generic Order message. Pandemonium will reign until such an effort emerges.

In terms of browser support, Version 5 of Microsoft's Internet Explorer browser will support XML, as will that from Netscape. These will also support XSL to describe how the received XML should be displayed. Whereas HTML is really limited to presenting for display, XML can be put to a much wider range of uses. In particular, it can be used to tag data for storage, or for transfer between applications. The Office 2000 suite from Microsoft will have an option to save in XML format, and it is rumoured that they intend to move to using XML as an 'open' interchange format, and use it to replace their own proprietary RTF (Rich Text Format) standard.

As indicated above, companies are developing document types to support the exchange of business related information. Some sets of document types that together can be used for a particular purpose have been given names by the companies or groups that have developed them. Here are some examples:

- Resource Description Framework (RDF)
- Open Software Description (OSD) format
- Distribution & Replication Protocol (DRP)
- Web Interface Definition Language (WIDL)
- Meta Content Framework (MCF)
- Channel Definition Format (CDF)

- Open Financial Exchange (OFE)
- Open Trading Protocol (OTP)
- MathML
- Chemical Markup Language (CML).

These various uses of XML make use of the XML syntax standard. As well as the basic syntax standard itself, there is a developing family of related new specifications, which include:

- eXtensible Markup Language
  —User defined data tagging
  —Transmitting of solution neutral structured data
- eXtensible Forms Description Language
  —Digital representation of complex forms
  —Integrated computations and input validation
  —Multiple overlapping digital signatures
- Synchronised Multimedia Integration Language
  —Scalable graphics
  —Multimedia presentations with linked audio
- Resource Description Framework
  —Better cataloguing of content
  —Search facilities, Content rating
- eXtensible Style Language
  —Control of document content
  —(CSS will style web documents)
- XML Schema
  —The XML schema language can be used to define, describe and catalogue XML vocabularies for classes of XML documents. The purpose of a schema is to define and describe a class of XML documents by using these constructs to constrain and document the meaning, usage and relationships of their constituent parts: datatypes, elements and their content, attributes and their values, entities and their contents and notations. Schema constructs may also provide for the specification of implicit information such as default values. Schemas document their own meaning, usage and function.
- Document Object Model
  —A platform- and language-neutral interface that will allow programs and scripts to dynamically access and update the content, structure and style of documents. The document can be further processed, and the results of that processing can be incorporated back into the presented page.

It is instructive to compare the current situation of XML with the historical precedent of EDI. The need and basic ideas for EDI emerged and were developed independently by several different groups. Today, we have the US using mainly X12, the UK using Tradacoms, the airline freight world using CARGOIMP, and there are several other incompatible variations. Eventually, the world got its act together and formed an international effort under the auspices of the UN to agree a single set of data definitions, out of which to build a single set of EDI messages—the whole thing being called EDIFACT. However, EDI has remained fragmented in practice.

The process being followed in developing XML is, in many respects, similar. A single internationally recognised consortium, W3C, is developing the syntax for the XML language. Trade groups have since emerged to provide data definitions and specialised 'messages' for their sectors. However, there is no body that is ensuring that duplicate groups do not form, and that groups are prevented from developing standards outside their defined areas.

What is missing is the all important middle layer. No international group has yet started to produce common XML data definitions and messages (such as a generic Order message) that can be commonly used and built on by others. Worse still, groups of companies are trying to commandeer this space for themselves. Other groupings may also emerge to muddy the waters yet further.

7.10 SUMMARY

For all of the hype about online shopping, it is supply chain automation that will really test the viability of eBusiness as a significant shift in the way that the world works. This chapter has exposed the issues that need to be addressed in implementing electronic dealings between suppliers, notably;

- Workflow, and how the implicit series of transaction in any activity need to be clearly, accurately and unambiguously defined if they are to be automated.
- Direct and indirect goods, and why it is necessary to distinguish between general purpose and specialist purchases.
- Multiple suppliers, and the need to distil a single view of supply options from multiple feeds.

In each case, the current options (such as market sites to cope with multiple suppliers) are explained and discussed. In addition, the support technologies for Supply chain automation, EDI and XML, are covered in some depth.

Perhaps the most important point made in this chapter is that implementing an electronic supply chain is a major and potentially irrevocable step. Links to legacy suppliers may remain, but any organisation intent on full-strength eBusiness will, in practice, need to fundamentally change its operating processes.

Hence, if the rewards of increased efficiency, speed and choice are to accrue, significant up-front commitment and investment is needed. Of course, the view of industry leaders is that supply chain automation is only one option—you don't have to survive!

FURTHER READING

Norris M, Davis R & Pengelly A, *Component-Based Network Systems Engineering*, Artech House, 1999.

Parfett M, *The EDI Implementors Handbook*, National Computing Centre (NCC), 1992.

UN/EDIFACT Standards Database, http://www.itu.int/itudoc/un/edicore.html

RosettaNet, http://www.rosettanet.org/

OBI, http://www.openbuy.org/obi/about/OBIbackgrounder.htm

8

Setting Up Shop

Opportunity is missed by most people because it is dressed in overalls and looks like work

Thomas Edison

For all of the technical gloss and wizardry behind eBusiness, there is no reason why it has to be difficult to implement. Of course, it helps to understand how all of the various components work together, but you can readily set up an online shop with little or no practical experience. This chapter is a brief guide to anyone taking the plunge. In keeping with the rest of the book, we focus on the technology and assume that the business case, aesthetics and market positioning have all been thought through.

To start, we explain the options that are available when setting up shop and the decisions that have to be taken. This is backed up with some pointers to the various players who can help and a checklist that goes, step-by-step, through building a viable eBusiness.

8.1 READY, STEADY . . . SHOP

Even the best of plans will founder if the basic equipment is not in place. To run an eBusiness, there is some equipment that you need to have access to. That doesn't mean that you have to own it or run it (in truth, there is very little you cannot borrow or hire)—you just need to have access. The basic kit includes a server, to host your 'shop', a catalogue to advertise you wares, some form of connection to a network (notably the Internet) and an interface to your existing systems. Over and above this, you should have some applications (or your shop will be rather boring) , a payment mechanism (if you want to make any money) and some decent security (if you want to have any credibility). For each of these component parts, there are some choices on offer and some decisions to make. So, let's look in a bit more detail at each part.

Server

As everyone knows, you need one of these to host your shop. The first question to ask about the server is whether you manage your own or whether you outsource the job to a third party. There are loads of companies who specialise in running servers, and they will happily lease you some space on their network-connected, maintained computer.

So, do you go straight for a hosting service? The answer in most instances is yes, but there may be instances when a better option looms. For instance, if the nature of your eBusiness is oriented more on business-to-business than consumer-to-business transactions, it could be good move to build up some technical expertise. In this case, buying a server and hiring a 'webmaster' with experience of server operations, applications and maintenance would be worthwhile.

In some organisations, there is already some in-house capability to handle full time (i.e. 24 hours a day, 7 days a week) computer operations. If such people are available, everything can be kept in-house, thus engendering a warm feeling of control for many people. The cost of engaging a third party is also saved, and this can offset the cost of keeping the local talent up to speed and strength.

There are pros, cons and myths whichever route is taken. The myth to dispel is that hosting services are for 'small' sites. Some of the top sites on the Internet use third parties, and more than half of all online business is transacted thanks to hosting services. There are clearly some instances when hosting is the only option, but all other things being equal, the choice between the two is really one of convenience and cost.

As with everything else on the Internet, eBusiness relies on co-operation and interworking, so having direct control over your own server doesn't guarantee overall control. Nonetheless, having some technical expertise on-hand may be useful if you need to do anything a little out of the ordinary.

Catalogue

There is not much point in setting up an online shop if you don't have a clear idea of what you are selling and to whom. You need a catalogue to advertise your wares, and this catalogue needs to be appropriate to the trading model that you are going to adopt.

We saw in Chapter 3 that there are different styles of catalogue depending on whether you are dealing with consumer-to-business or business-to-business. If it is the former, then style and design are important. When a customer can be lost or won in seconds, goods and services must be presented to appeal, and it must be easy to navigate the catalogue. Furthermore, it must be easy to complete a transaction. The general guidelines is that the easiest way to sell anything is to make it easy to buy it.

The rigours of getting a good business-to-business catalogue are somewhat different. Design flair matters less, but accuracy is paramount. Because this sort of catalogue tends to be fairly extensive and used more by computers than people, some process for managing its content is called for. A simple validation process which ensures that entries are not misplaced, lost or incorrectly entered should be established here.

Network connectivity

There are many different options for hooking your server up to the rest of the world. At the very simplest level, it can be connected using a modem and an ordinary telephone line. This works, but sets a very low limit on the number of visitors to your shop—a good number will just go away, put off by congestion, slow response times or simple unavailability. So, it is worth thinking about what sort of access you need to provide to your shop, and this is a balance of expense against service.

One step up from the plain, ordinary telephone service is the Integrated Service Digital Network (ISDN). This looks much the same a normal phone line, but provides about twice as much bandwidth. Transactions are that much faster, but neither ISDN nor standard telephones are really intended to provide 24×7 access to an online server. They are both switched networks, and to get lots of people into your shop anytime of day or night, you really want a dedicated line.

There are several options when it comes to high speed access—data networks such as frame relay and ATM (Asynchronous Transfer Mode) can give assured bandwidth between your server and the public Internet, and leased lines (T1 and E1 links) can be used to connect to a router, which then connects to the Internet. All of these options operate in the Megabits per second region, as opposed to the Kilobits per second offered by the switched network services. It almost goes without saying that most of the top Internet sites have opted for a high speed link.

As well as having a public connection to your own server/shop front, the connection to other (private) networks may be important. This is especially true for business-to-business trading, where suppliers and customers often have their own Virtual Private Networks (VPNs), Internets or Community Of Interest Networks (COINs). The issue of having a high enough bandwidth connection is the same, and it is here that dedicated T1 and E1 links are a good option.

Interfaces and integration

In all but the very simplest eBusiness configurations, there are some interfaces to existing systems. These may be very simple interfaces (such as a fax machine or a swivel chair with a very busy clerk on it), but if there is any intent to keep stock control, deliver goods, maintain a general ledger or issue bills, back-end systems need to be integrated.

The technical side of integration can be quite complex, as was shown in Chapter 6. Also, practical integration of systems is a messy business that can sometimes call for quite a lot of technical expertise. Having said that, there is a lot that can done without recourse to a Unix guru or a database whizzkid; there are toolkits for building a straightforward shop (e.g. EC builder from Multiactive) and services that handle your online business and email you when an order comes in (e.g. see www.StoreCentre.bt.com).

At a basic level, the flow of orders, bills and enquiries should be traced through the shop. A simple walk-though of all of the key processes will identify where information is used, processed and stored. From this, it is fairly straightforward to see where automation will work, where it won't, and where some form of manual work-around needs to be put in place.

A fully automated solution may or may not be required—again, this is a decision that should be taken as part of an initial business case. In any instance, it is worth following the main trading processes through, as it is an easy and low cost way of highlighting where integration work may be required.

Applications

The term 'applications' is used by many in the computing business to cover just about everything that isn't an operating system or database. When it comes to eBusiness, the most basic application is something that allows orders to be taken. Not much further on in terms of sophistication are applications that support invoicing, taking credit card details, logging transactions and tracking customers. As we've already seen, commercial packages can be used in many cases, but when some extra customisation or flexibility is needed, a special purpose application has to be built. So we look now at a few of the environments that can be used to build these applications.

Development systems for eBusiness come in all sizes and shapes, from the sophisticated products aimed at big businesses down to the simpler (and cheaper) toolkits intended for the more general market.

At the high end, two of the main offerings are Open Market's LiveCommerce and Netscape's ECXPert. Both of these products offer a full range of facilities and speed the development of all manner of server applications. Although very flexible in terms of what they enable, both are expensive (tens of thousands of dollars) and restricted to specific operating systems (NT, Solaris).

More modest packages are on offer from Microsoft (Site Server Commerce) and IBM (Net.Commerce). Both of these can be regarded as toolkits rather than full eBusiness development environments, as they tend to assume specific market models. That said, they are reasonably priced and tend to run on a range of operating systems.

In addition to products badged as eBusiness tools, many of the established computer suppliers have products that can be used to develop eBusiness applications. For instance, the database company Oracle supplies a comprehensive range of tools for integrating databases with the World-Wide Web.

Security and payments

In most instances, the main concern for an online business is to protect credit card payments. This is a specialised area that all but the very brave would tackle without help. Fortunately, it is also one that all online merchants have to deal with, and so one where there are a whole host of established suppliers who can provide you with a ready made payments package. Typically, these suppliers provide software that, once loaded onto your server, assumes the familiar role of a merchant's cash register or point-of-sales terminal.

The main functions that should be provided in any of the commercial payment systems are the authorisation of transactions, the provision of electronic receipts and the handling of returns and receipts. In addition to these basic secure credit card functions, there may be provision for manual card processing and administrative support for checking transactions, audit and accounts. These additional feature become important if there are no existing back-end administration systems.

As well as acquiring a commercial package, there are a couple of other things that need to be done in setting up a payments scheme. The first is to contact a bank and set yourself up as an authorised merchant. As explained in Chapter 4, the bank needs to authorise transactions and carry out the final settlement. So they need to be included, and preferably at an early stage—banks can help in selecting the settlement arrangements that best suit the business you are in (e.g. low value items paid for with electronic cash, high value items paid for with credit cards, etc.).

Once a bank has certified you as a merchant, they will issue an identifier and establish this with a credit card processor. This identifier is has to be entered into the payments system before you can begin.

The broader area of network security can be quite a complex one—witness the small and highly prized teams of security experts that most large organisations guard. For the normal person, many of the key security issues are left to be dealt with by a third party—firewalls and network security can be provided by a hosting service, encryption and authentication by a trusted third party, and secure payment by merchant service.

The bolt on goodies

Just as you can set up an online shop with virtually no equipment, you can go to the other extreme and deploy the full weight of modern technology to support your eBusiness. Most of our 'bolt on goodies' tend not to be available as commercial offerings, and so have to be developed specially. The best advice in how to acquire these specialised pieces of software is to ensure that good development practices are followed. In the context of eBusiness, the main areas in which technical development takes place are:

- Those that are concerned with additional application functionality such as integration with stock control and the automation of payments. In this instance, you either have to carry out some development on your applications platform or pay someone to do it for you.
- Those that can be configured onto the existing technical configuration (without further development) such as enhanced security which would be effected by, for example, the use of server certificates and secure sockets.

In addition to these, there are some 'bolt on goodies' that are concerned more with the network infrastructure than applications software. The prime contender here is a network with an assured quality of service, something that is usually available, at a price, from a network provider. Having a guaranteed access bandwidth ensures that your customers have an unhampered way of getting at your wares—it is rather like using a security firm to remove buskers from outside the shop entrance.

8.2 WHERE YOU GET YOUR STUFF

There are many choices of suppliers in eBusiness technology, and even more pieces of kit that you can purchase. Some of the suppliers offer a complete service and some specialise in a particular area.

Given the speed of innovation, it would be foolish to commit a lot of product information to paper, but there a core of players who, for one reason or another, seem destined to play a part in eBusiness both now and into the future. In this section, we will introduce the main categories of player, and outline roughly what drives them and what they have to offer.

Internet Service Providers

These are the people who started out by providing a point of presence on the Internet with, perhaps, some information content as well. Compuserve and America On-Line are two of the bigger and better known ISPs. As established companies, they are increasingly offering web space rental, application management and design services. With eBusiness and the Internet inextricably linked, they are well placed to support the online merchant, as well as the section of the general public who are most disposed to trading over the net. The extent to which their experience of serving a mass market with inexpensive service/products stands the eBusiness challenges remains to be seen.

Telcos and computer suppliers

The convergence of computers and communications is one of those phrases that has real meaning when it comes to eBusiness. Nowhere is this clearer than when we see how quickly both the established computer companies (Sun, Compaq and IBM) and the Telcos (BT, AT&T) have dived in. Both groups see eBusiness as the main growth area for their core products, and the Telcos, in particular, see it as a means of expansion in the face of a saturated telephony market. Both have a strong track record with premium services and volume markets, and both have significant depth of technical expertise. How flexible and cost effective they will be has yet to emerge.

Systems integration houses

It is probably abundantly clear from earlier chapters that there are significant technical complexities that underpin the magic. When expertise in software, networks, security, integration and project management are all needed to produce a customised solution, consultancy companies come to the fore. Andersens, EDS and the like have the breadth of expertise that enables them to deal with the more exotic and more complex eBusiness challenges. Needless to say, such expertise does not come cheap.

E-business specialists

In addition to the smaller scale consultancies (who will often do you a good deal) that cover all aspects of online trade, there are a number suppliers who provide a specific component of the overall solution. For example, there are several suppliers of payment systems—First Virtual, Digicash and Cybercash being some of the main contenders. As with any packaged solution, the judgement of applicability rests with issues such as flexibility (universal acceptance matters for any payment scheme, so the number of people are using a particular one is an important factor).

8.3 SHOPFITTING

Irrespective of whether you set up shop using third parties or handle the whole thing yourself, there is a significant element of project management that needs to be applied to ensure that things turn out as they should. Because we are looking at technical projects that involve (mostly) software development, integration, installation and test, it is worth looking at best practice in this area.

It is often said that good management comes from experience, and experience comes from bad management. The pain of getting things wrong can often be overcome by borrowing other people's experience. The idea of analysing trends and using them to build a framework for a job is the subject of much study in the computing fraternity, and rejoices under the name of 'patterns'. For now we'll go with the academic flow, but think of it as structured common-sense.

If we take an average across a number of technical projects, there is a typical distribution of time. To start with, there is a considerable proportion spent finding out what is really wanted. This is followed by a development phase, where the wants are engineered into a tangible product. Finally, we have an installation phase where the product is adjusted to meet the needs of its users (as opposed to the wants of its customer). This is illustrated in Figure 8.1.

This pattern seems to apply very well to a wide variety of projects. Indeed, it is largely based on network software development—the very stuff of eBusiness solutions. So it is probably worth looking inside each of the three phases.

The 'requirements' phase is all about deciding what the project is all about. Although nominally part of a technical development project, much of this phase is really about understanding the business proposition, sorting out politics, and even just waiting for inspiration—in the dynamic eBusiness environment, you have to do something, even if it is not clear what that something is! All too often, the initial proposition is that 'something must be done, and this is something, therefore this must be done'. It should be no surprise that it takes a while before the qualitative and quantitative details that a developer needs are available.

Once there are some hard requirements, the solution design can begin. Development lifecycles that systematically move through specification, logical and physical design and on to test and release are well established. In many instances, there are established metrics for development along with clear quality gates.

The third portion of the application arrow—known as 'customer engineering'—is less well controlled. It deals with adjustment in the light of user acceptance. The reason that this phase is unpredictable is that the customer or developer of any application rarely knows what an end user will do with it. Sometimes there are genuine problems with the product (it simply doesn't work); sometimes it works, but not well enough; and sometimes it

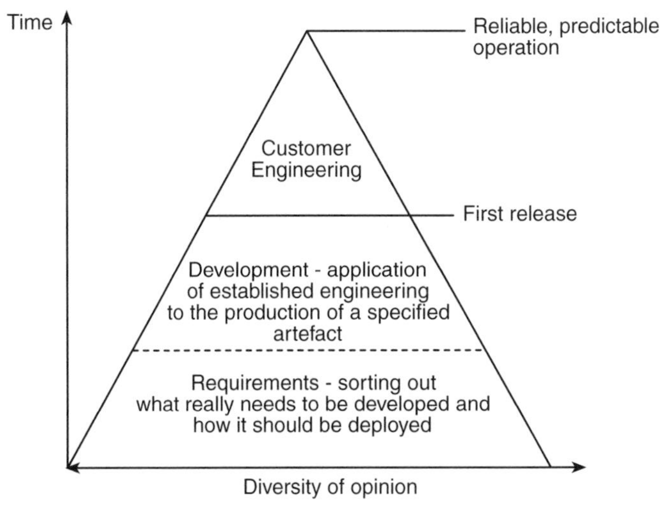

Figure 8.1 Where project time goes—the applications arrow

works fine but doesn't appeal. Hence, there are several strands within this phase. In addition to the well recognised repair aspect of post-release support, there are also preventive measures (keeping in tune with a changing world), perfective measures (making it slicker) and optimising measures (increasing its appeal).

For all their usefulness, established patterns and experience don't provide all the answers. They are a pretty good place to start, but the world moves on and, despite there being a similar picture of project time distribution for eBusiness projects, there is now a very distinct skew. The new version of our applications arrow is shown in Figure 8.2.

We have already hinted at why the development phase has been truncated—the raft of tools and packages that can be used to implement an eBusiness solution can make for very rapid delivery. There are a couple of other reasons. The first is that technical development is open to analysis and study. Because of this, a lot of attention has been focused on how to do it better and faster. So, as well as lots of ready-made software, the development community also has lots of ready-made know-how, encapsulated in design methods, quality systems and generic processes.

The second reason for speedy development is that the world of computers and communications has become more open over the years. That is to say that, for all the diversity of products, there is some measure of standardisation across the component parts of a system. The designer is working with well understood and (to some extent) interchangeable components that have predictable behaviour. So a solution can be construct more

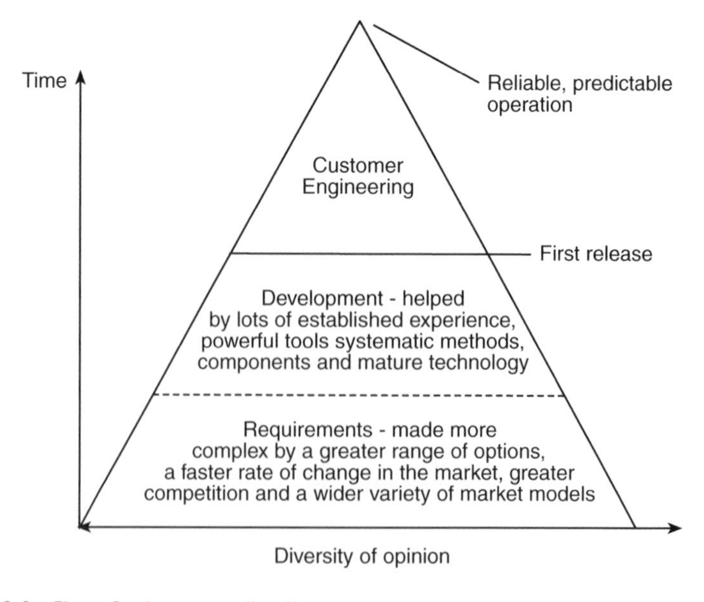

Figure 8.2 The eBusiness application arrow

readily than in the bad old days, when everything not developed in-house could be something of an unknown quantity.

Given this, the challenge in getting an eBusiness product to market is not a technical one (astonishingly). The authors would argue (and frequently do) that the real danger is spending so much time deciding what you are going to build that the opportunity passes. There is a lemma of West's Law here—if you take too long getting a product online, someone else will have cornered the market and the only entry will be to offer it for free.

At the other end of our deployment arrow there lurks another risk: if you put too much into support, you are left with a high maintenance product just at the time when everyone else is looking for a free solution.

After years of project management and technical development, our summary for the world of eBusiness is that if the engineering is the bottleneck, then you are doing it wrong. Business acumen rules, technology serves.

8.4 CHECKLIST

There a couple of points that should be clear from this chapter. The first is that there is a fairly logical sequence of questions that should be asked and decisions that should be taken in setting up shop. Most of these questions and decisions are straightforward—a mix of common-sense and basic know-how. The second point is that there are lots of specialists who can help in setting up an eBusiness. You don't need to own a server, or arrange for goods to be delivered, or establish a secure payment mechanism. You can do all this, if you want, but there is the option of using third parties. In fact, you can do just about everything using other people's equipment and services.

This section brings the two points together to present a checklist for setting up shop. We have covered most of the detail that lies behind the scenes already. The important thing now is to ensure that the transition to online trade is systematic, and for the business and technology that supports it to be in balance. So, the six steps are:

1. Decide what business you want to be in. Who are the customers, the suppliers, and what is your relationship with each of them? Adopt the appropriate market model from those explained in Chapter 2.

   The customers of the toy soldier supplier Trophy miniatures are the general public, and their main suppliers are metal and paint wholesalers. The goods that Trophy buy in are direct in both instances and, being a small concern, they have no indirect supplies worth speaking of.

   By way of contrast, the suppliers for the supermarket chain Tesco range from large manufacturers such as Heinz, Birds-Eye and Coca

Cola right down to market gardeners and small specialist suppliers such as Copella cider and strawberry growing co-operatives. The customers are still the general public, albeit in larger numbers than for Trophy, and they do have a significant requirements for indirect goods, as any other company—buckets, mops, cash tills and PCs.

2. Decide which aspects of the value chain are most applicable, the two extremes being cost cutting or revenue expansion, for eBusiness to address. The most important issue may be to reduce order handling costs (which is where Cisco saved money—see Appendix 1) or to access a wider market (which is where Trophy miniatures fits), or something else.
3. Get a web presence (if not you don't already have one). The issues covered earlier in the chapter come into play here. The main decision—whether you are buying into a hosted platform or have your own set up—carries significant pros and cons:

	Pro	Con
DIY	Control, Flexibility, Integration with other systems, processes.	Cost (particularly costs for your own internet feed—router, firewall). Stuck with legacy, Need own IT skills.
Third party	Range of expertise provided (Page design, Security, etc.). No capital costs.	Cost (bespoke service may attract a premium). Perceived loss of control.

4. Decide how to present your goods. This entails establishing the right sort of catalogue for the job, as explained earlier, as well as deciding what sort of image and style to project.
5. Sort out order taking and fulfilment. The former implies (at least) presenting the customer with an order form and, more likely, entails establishing links to back-end systems for stock control. The latter applies for any physical goods—the deal may be struck online but real products have to be delivered to real front doors! The preferred solution for fulfilment would be to engage one of the logistic companies (UPS, FedEx) by integrating with their systems (a task they have made easy—see Appendix 1).
6. Establish a payment scheme. Unless you are a charity or very philanthropic, you'll want to earn some cash from your online investment, so a mechanism for receiving payment is essential.

There is more that could be done, but at this point you are ready to do business. Of course, the success of the business relies much more on its positioning, presentation and differentiation than the technology, so this should be seen only as the end of a beginning, rather than any sort of end in its own right. But any idea transplanted in a new place tends to spark innovation, and it seems likely that the migration of business onto a network will do likewise.

8.5 SUMMARY

This chapter has dealt with the nuts and bolts issues of setting up shop. We have presented the main options that have to considered in eBusiness—whether to buy or rent a server, where and how to use third parties, and what sort of technology is available. The pros, cons and (sometimes) myths that should be considered in making decision on these options are explained and illustrated in a six-point checklist at the end of the chapter, which recommends:

1. Deciding what business you are in.
2. Deciding where eBusiness offers advantage.
3. Choosing the appropriate technical options.
4. Choosing the on-line style that you want.
5. Sorting out order handling and fulfilment.
6. Sorting out a payment mechanism.

The main messages in this chapter are that all decisions should be driven from business need, that a systematic process should be followed, and that setting up shop is only a start—it is what you do with it that really makes a difference.

FURTHER READING

Yesil M, *Creating the Virtual Store: Taking your web site from browsing to buying*, John Wiley & Sons, 1997.
Kerth N & Cunningham W, *Using pattern to improve our architectural vision*, *IEEE Software*, January/February 1997.
Norris M, Davis R & Pengelly A, *Component-Based Network Systems Development*, Artech House, 1999.
Norris M, *Survival in the Software Jungle*, Artech House, 1995.
Orfali R & Harker D, *Client/Server Programming with Java and CORBA*, John Wiley & Sons, 1997.

All of the suppliers cited in the chapter have web sites that provide a lot more detail on their ideas, products and services.

9

Underlying Technologies and Standards

Technology is dominated by two types of people: those who understand what they do not manage, and those who manage what they do not understand

Putts Law

Many years ago, we were handed the career opportunity of becoming software maintainers. In seeking clarification from our new boss as to the nature of the job, we were told that it was all about keeping established, software-rich products in tune with customer requirements. So we would be required to optimise what was already there to make it run faster or better, identify and fix problems as they arose, and add any new functions that were needed to keep the products fresh. But what sort of training or preparation should we do for this new challenge, we asked. We were informed that our question was a bit of a tough one, but a good maintainer had to know a bit about everything.

Despite being rather depressed at the time over this need to become omniscient, we have subsequently discovered that a broad technical awareness is invaluable to many jobs. An enduring learning point from our time in software maintenance was that many are bright enough to explore a topic—the trick is to know that it is there to be explored in the first place.

Although little more than thirty years old, the range of computing and information technology is vast. No-one can really grasp all of it in any sort of detail, yet there are a few ideas that will sustain you through most challenges. eBusiness brings together a whole raft of different devices, networks, concepts and standards. We have already gone into considerable detail on security, payment and other issues that are central to online markets. In this chapter we provide a minimal overview of all of the other elements that play a part.

There are plenty of detailed guides to each of the topics addressed here, but first you have to know what is there and what it can do for you. In the words of Sir Isaac Newton, 'if we see far, it is only because we stand on giants shoulders'. Despite being a mere youth in engineering terms, there have been quite a few giants, and their legacy is there for all to use.

9.1 ACCESS DEVICES

It seems increasingly likely that the standard desktop PC will be but one of a number of eBusiness interfaces. Already there are several alternative access devices being used, piloted or on trial. Here we look at some of the main alternatives to the PC.

Internet screenphones

Internet screenphone (or web phone) products have been announced by various vendors. These devices combine the facilities of a telephone with an Internet access device, by providing a handset and keypad, a liquid crystal display touch screen (typically with VGA resolution), a small profile keyboard for data entry, and a smart card slot for secure access. They contain a processor plus RAM, with ROM for basic software and flash RAM for updateable software and configuration data. Various options for Internet connection are possible, including PSTN via modem, ISDN and, in the longer term, ADSL. These network technologies are explained in Section 9.3.

The devices provide advanced telephony facilities, including integral directories with direct dialling, incoming and outgoing call logs, and user-friendly, screen-based interfaces to network services such as call waiting, call diversion and three-way calling. In addition, they will provide full Internet access, including the Web and the potential for convenient unified messaging, combining access to voice, fax and email messages.

These devices are clearly aimed at the consumer and small business markets, but they also have the potential in specific areas of corporate markets to combine the functions of the existing desktop PC and telephone.

Set-top boxes

The Set-Top Box (STB) is a peripheral device that connects to a standard TV set and transcodes incoming signals for display on the TV. STBs are used to connect to cable and satellite services, and more recently to digital TV services via satellite or terrestrial signals. For interactive services, the connection from the user to the service is typically provided via a modem in the STB connected to a phone line.

These types of two-way STBs provide access to video and multimedia services, and e-commerce services such as home shopping and home banking. The user input device can be a standard remote control, or possibly a keyboard connected by infrared link to the STB. User access security can be provided by smart cards, which could also be used for e-commerce applications, such as loading cash cards within home banking applications.

STBs were first introduced way back in the 1970s with the emergence of cable TV companies in the US, providing numerous TV channels. The cable TV providers defined the early STB specifications, but these were mainly proprietary rather than agreed standards. However, there has been some government standardisation effort in the US.

A recent innovation is an STB, often called a net-top box, which provides access to the Internet on the TV. Based on a subscription-based service, a proprietary browser is provided which enables Web pages to be viewed in a similar way to a PC, but with some adaptation of the text and images for the limited capabilities of the TV display. Email and other applications can also be used, but as there is no local storage in the TV, data has to be stored with the service provider or in the STB itself.

The battle for dominance of the STB vendor market is being strongly influenced by the choice of Operating System (OS). Web TV, now owned by Microsoft, uses Windows CE, whereas other vendors such as TCI and Hitachi are looking to use the Personal Java OS. Sun, the developers of Java technology, have now released the Java TV API to encourage further Java developments in the STB market. Internet TV services are available in parts of the US, and (if they follow the usual pattern of technology diffusion) are likely to appear in the wider world over the next few years. See Section 9.5 for a little more on operating systems for access devices.

Kiosks

Multimedia public kiosks offering Internet access in the high street are relatively new devices. They can offer users convenient access to e-commerce services when they are out and about, without the need to carry personal devices. Just like cash dispensers, they tend to be strategically located in places where users are likely to require their services, for example at airports for car hire, hotels for booking travel and theatre tickets, and shopping centres for banking services. There are two basic types of kiosk:

1. Internet based kiosks—these are kiosks that offer specific Web information and services via a commercial wrapper, hiding the complexities of traditional PC Web browsers.
2. Web payphones—provide pay-as-you-go Internet access, for use of email and the Web. Also known as public Internet terminals or Internet payphones, they could replace existing pay phones in suitable locations.

Users of kiosks get up-to-date Web information, and can be put immediately in contact with any of the advertisers' sales desks, and make payments via credit card. Print-outs can also be obtained on some devices. The service is usually free to the user, funded by advertisers who are being provided with new delivery channels.

Mobile devices

There are now a whole host of handheld devices that can interact with eBusiness applications. Some of the larger devices, such as laptop PCs, connect to a network using standard PCMCIA modem cards, but for smaller devices, such as palmtop computers, it can be more complicated. For example, to connect a Palm III to a GSM network requires a 'Snap-On' modem, an interface adapter, a GSM phone and connecting cables.

Infra-red connection between devices and mobile phones, using the Infrared Data Association (IrDA) protocol, can simplify this complexity, but requires line of sight association between devices. Short-range radio link connections between devices using unrestricted 'free band' radio frequencies will soon provide a less restricting alternative for portable devices

A now familiar alternative to adding connectivity to computing devices is to add computing functionality to a communication device. The Nokia 9110 communicator exemplifies this approach, and provides email, fax, short messaging, web access and office capabilities in a palmtop device. The need to have a reasonable sized screen and keyboard constrains the minimum practical size for such a device, but further innovations are likely to address these issues.

Much research is going on into wearable interface devices and the capabilities they would provide. For instance, Swatch are currently building GSM capability into their watches. More exotic options are monocular and head-mounted displays, speech driven interfaces and a glove-based input device. Although somewhat outlandish, they may well form the basis of practical future devices.

Figure 9.1 The Easi-Order terminal

In addition to these fairly generalised devices, there are likely to be some very specific eBusiness access products. With device costs falling, enterprises such as retailers and banks will produce special-purpose devices specifically for use with their services. By way of an example, the UK supermarket chain Safeway have started an electronic shopping trial called Easi-Order, with 200 shoppers being issued with commercial versions of the 3Com Palm Personal Digital Assistant (PDA), with the addition of a barcode scanner in the hood (see Figure 9.1).

A customised store download is available to each shopper's PDA, based on individual shopping preferences. New items can be chosen and scanned in for immediate purchase, and are added to their profile to appear in their next downloaded list.

With so many different ways of initiating an electronic transaction, there has to be some way of allowing a user to transport their 'personality' (i.e. their identity, cash, access privileges, etc.) from one point of access to another. This is where the smart card seems destined to play a role, so we move on to look at these in some detail.

9.2 SMART CARDS

The smart card is a credit card-sized plastic strip with some integral memory and processing capacity (Figure 9.2). It has external connectors but doesn't usually have keys or buttons like a calculator. It can be used as

Figure 9.2 A typical smart card

an electronic purse or a multi-function security pass. Smart cards can have any permutation of processing power, memory and security features. Current technology is capable of producing cards that can carry out the complex mathematical calculations to encode data so that it can be safely transferred over networks. The memory on a card can be reprogrammed around a million times, failure rates are virtually negligible, and the cost of the smart card is comparable to that of a phonecard.

Features such as card size, contact layout and electrical characteristics have been defined by ISO (International Standards Organisation) standard 7816. At the functional level, smart cards can be categorised as either memory cards or microprocessor cards. Memory cards, such as disposable pre-paid payphone cards or loyalty cards, are the most basic and cheapest form of smart cards. They contain a small amount of memory in the form of ROM (Read Only Memory) and EEPROM (Electrically Erasable Programmable Read Only Memory). The capacity may only be a few tens of bytes, but the 'non-volatile' nature of the memory enables the storage of simple data, such as decrementing value counters. This can be retained after the electrical power supply provided by the terminal is removed. A useful feature as, smart cards do not have their own power supply.

Microprocessor cards are more advanced than simple memory cards in that they contain a microprocessor CPU (Central Processing Unit) and RAM (Random Access Memory) in addition to ROM and EEPROM. The ROM contains the card's operating system and factory-loaded applications. Smart card operating systems tend to be proprietary, and there is as yet no widely accepted standard operating system, although Multos and Smart Cards for Windows will soon be vying for that title.

Although ISO 7816 is the fundamental standard for smart cards, it tends to focus mainly on interoperability at the physical, electrical and data-link protocol levels. That said, it does also define a standardised logical structure and a small set of commands, but this was adopted as an international standard by ISO only relatively recently. This has meant that the smart card market has evolved on a largely proprietary basis.

With no widely accepted standard smart card operating system, card manufacturers have tended to develop their own unique operating systems, specific to the particular chip hardware. Card applications are then specific to a particular operating system embedded in the chip. Thus, even in schemes like GSM, where cards from different suppliers are interchangeable, the firmware in them is proprietary and non-portable.

Smart cards have tended to hold only one application per card, hence users have had to carry several cards to meet their needs, typically from different manufacturers. The fact that end users need an extra piece of hardware, the smart card reader, is one obstacle. Requiring them to use more than one manufacturer's reader is another. We have yet to reach the stage where desktop computers are either supplied with or can be fitted with industry-standard smart card readers in which smart cards for a variety of different applications can be used interchangeably.

In 1996 Europay, MasterCard and Visa (EMV) defined a specification that extended ISO 7816 with the addition of data types and encoding rules for use by the financial services industry. The European telecommunications industry also embraced the ISO 7816 standards for their Global System for Mobile (GSM) communications smart card specification to enable identification and authentication of mobile phone users. For PC-based smart card applications, further (*de facto*) standards have emerged to address the fact that there was no standard way for PCs to talk to smart card readers, with users being locked into proprietary solutions where a particular manufacturer's card could only be used with that manufacturer's reader.

The PC/SC (Personal Computer/Smart Card) work group was formed in May 1996, backed by some major players in the PC and smart card markets such as Microsoft, HP, IBM, Sun, Gemplus and Schlumberger. It was created to develop open specifications for a standard model to interface smart card readers and cards with PCs. The PC/SC specifications are based on the ISO 7816 standards, and are compatible with both the EMV and GSM industry-specific specifications. They provide a layered model, shown in Figure 9.3 that allows smart card-aware applications to talk to the smart card in a transparent manner, in much the same way as applications today talk to other peripherals such as printers.

At the topmost level in the model, developers can write smart card-aware applications in a device (card reader) independent manner. Below this,

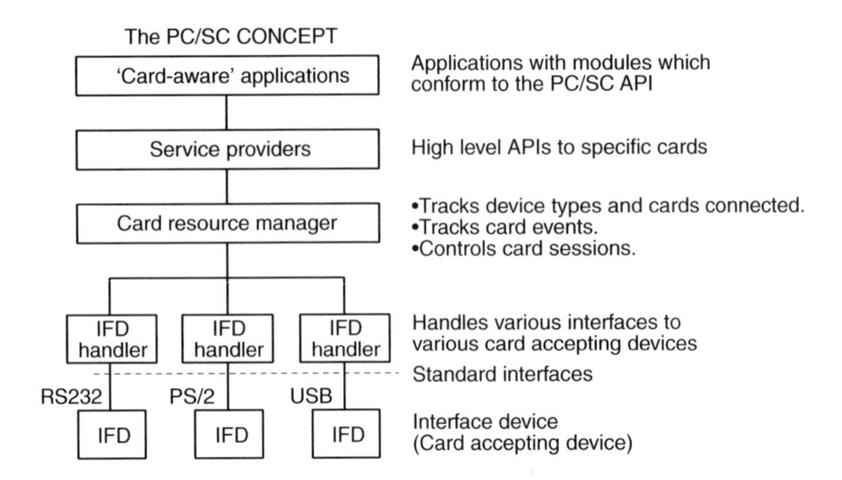

Figure 9.3 The personal computer/smart card model

the smart card manufacturers provide high level APIs (Application Programming Interfaces) for specific card functions. At the lowest level, and transparent to anything above it, manufacturers provide PC/SC compatible smart card readers and device drivers for the particular Windows platform.

Two interesting new smart card technologies that address the issue of lengthy development times for new applications are Java Card (which is an API) and Multos (which is an operating system and API).

Java Card

A Java Card is a smart card capable of running programs written in a much-reduced version of Sun's Java programming language. It is a specialised implementation since smart cards impose many constraints, such as the use of 8-bit integers instead of the 32-bits normally required. The Virtual Machine (VM) that translates Java byte code into card processor specific instructions, language specification and core packages all have to be made more compact for smart cards.

The Java Card API defines the calling conventions by which an applet accesses the Java Card runtime environment and native services. This allows applications written for one Java Card-enabled platform to run on any other Java Card-enabled platform.

On a Java Card, applets or 'cardlets' sit on top of the Java VM, which in turn sits on top of the card's operating system. The applets are stored in EEPROM, hence the ability to download new applets and delete unwanted applets during the card's lifetime. This provides great flexibility in that multiple applications can run on a single card, applications can be changed or corrected without having to issue a new card and the time to market for new applications is greatly reduced. The major disadvantage is that the hitherto fixed and tamperproof features of smart cards are now potentially opened up. Malicious applets could be unknowingly loaded, causing increased security concerns. In addition, the speed of execution is a problem for Java-based programs due to the extra layer of activity introduced by the Java VM.

Some Java Card products are currently available or in production, such as Gemplus' GemXpresso, Schlumberger's Cyberflex and Bull's Odyssey, but it could take some time for Java Cards to reach maturity. The Visa Open Platform specification aims to supply extra features, including aspects of security, which are not currently addressed by the Java Card API.

Multos

An alternative and more mature technology for multi-application smart cards is Multos, promoted by the Maosco consortium. Set up in 1997, this group includes Multos founders Mondex International amongst other industry leaders. The Multos specification is more comprehensive than the Java Card specification, defining the physical chip security and operating system as well as the virtual machine and APIs for programming it. Like the Java Card API, Multos is compliant with the ISO 7816 and EMV standards. Applications are again stored in EEPROM, but Multos provides a certification scheme for the lifecycle of applications: secure downloading, storage and deletion. Multos checks the validity of new applications being downloaded, and provides an area of memory for each to reside in. The design of the system provides a logical 'firewall' around each of the applications that will stop any interference between them. Each application is interpreted and talks to external systems via the operating system. This clear separation of applications on the card is vital. Some companies may only be happy having their application added to a multi-application card if they have some confidence that other rival applications on the same card will not interact with it. Maosco claims to be aiming for ITSEC certification level E6 for type-approved Multos cards.

Applications for Multos smart cards can be developed either in a low-level language called MEL (Multos Executable Language) or in a high-level language such as C. MEL is part of the Multos specification, and is a language optimised for 8-bit smart card microprocessors and the Multos-specific security requirements.

Applications

The smart card is capable of providing secure access to a variety of services, independent of delivery channel. Some of the most common uses for smart cards today fall into this category. Satellite television decoder cards are one example, along with SIM (Subscriber Identity Module) cards that authenticate GSM mobile phone users. There are many more besides, particularly as we move into an age of network computing, kiosk type terminals and all of the other access devices mentioned in the first section.

A mechanism often employed to ensure that the smart card is being used by its authorised user is known as Card Holder Verification (CHV). Commonly this is achieved by the off-line entry of a PIN known only to the card and its rightful holder. A correct entry will authenticate the card holder, whereas a successive number of incorrect attempts will lock the card (or an application on it) against further attempts. Authentication with a smart card is an example of 'two-factor' authentication, requiring 'something you own' (the smart card) and 'something you know' (the PIN).

Payment is a key application for smart cards. In 1996, Europay, MasterCard and Visa jointly published a specification called 'IC Card Specifications for Payment Systems'. This specification became known as EMV, and covers design aspects for smart cards, terminals and debit/credit applications used in financial transactions. One of the drivers for the move to smart cards is the ease and low cost in which magnetic stripe credit/debit cards can be forged, causing increasing fraud losses. A measure of the success of EMV is that many smart card products are now advertised as 'EMV-compatible'.

The EMV specifications were designed such that card terminals are not required to contain Secure Application Modules (SAMs) for authenticating cards. This is due to the high cost of managing the worldwide distribution and update of symmetrical keys to partly off-line card terminals. Instead, EMV uses asymmetric techniques to authenticate the card, adopting either static or dynamic data authentication. These are both based on digital signature technology, meaning the terminal will not need to contain any private keys.

There are some outstanding issues, and these include transaction speeds, staff training and the integration of Point Of Sale (POS) terminals. In response to these, APACS have announced that EMV-compliant chip cards will replace magnetic stripe credit cards in the UK over the course of the next two years. Furthermore, APACS has set the target that, by April 2002, 65% of all credit and debit transactions in the UK will use smart cards.

Currently, many users of public key cryptography store their private keys on their desktop computers. This private key database tends to take the form of an encrypted file on the hard drive, with access being controlled by a single password. There are many risks associated with this practice.

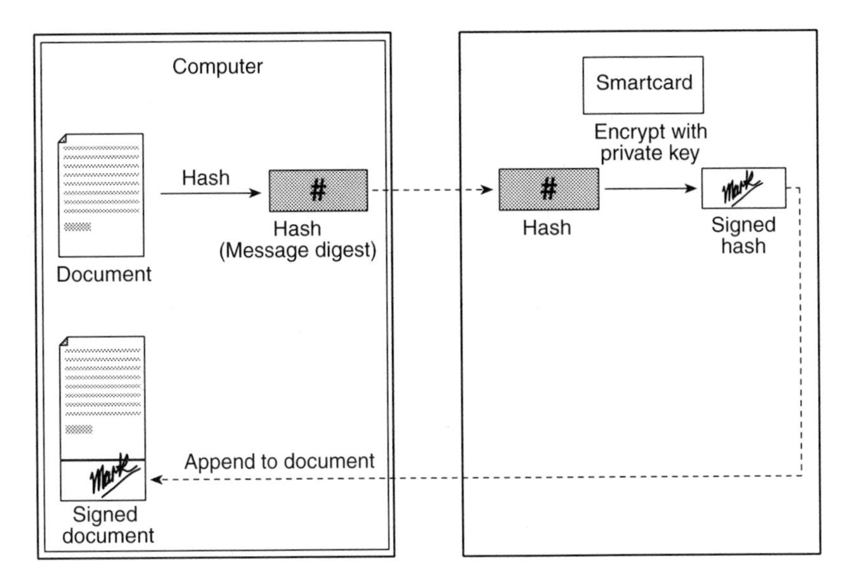

Figure 9.4 Signing a document using a smart card

How well the keys are protected ultimately depends upon the strength of the password chosen by the user.

Smart cards can be used to address these security weaknesses, with varying degrees of improvement that can be chosen to match the trade-off between perceived risk and cost. Security is of paramount importance in the design of an operating system for smart cards, something that is not always the case for desktop computers. The first generation digital signature cards simply provide secure private key storage. The card still relies on some form of PIN or password authentication, but is designed to be more tamper resistant, both physically and by dedicated software. While this results in a secure and portable storage medium for private keys, the cryptographic processing is still provided by the host computer, hence private keys are still read from the smart card for operations such as digitally signing and decrypting messages. This is illustrated in Figure 9.4.

9.3 INTERNET TECHNOLOGY

The Internet is a communications network that spans the globe. It has grown through the interconnection of many local area and regional networks, many of which are built on top of the data networks that we look at in the next section. The hierarchy of networks that form the Internet is shown in Figure 9.5.

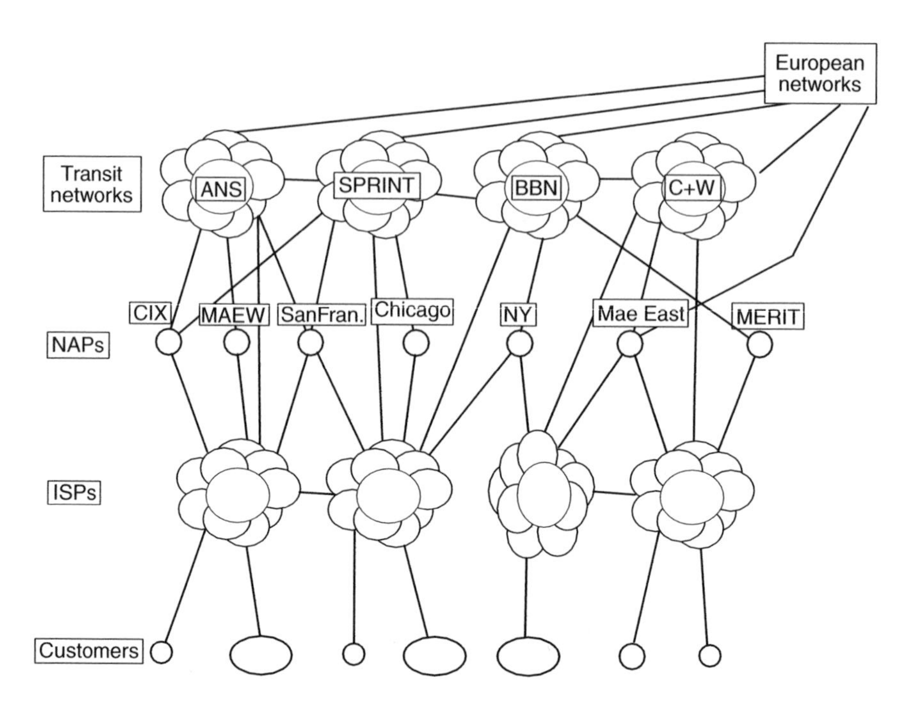

Figure 9.5 The network of network that form the Internet

The thing that gives the Internet its universal appeal is that it works by passing data with a standard and very widely implemented communications protocol known as the IP (Internet Protocol). The IP protocol can be used on many types of computer and over almost any network infrastructure: local and wide area. The Internet has a common naming and addressing scheme known as the Domain Name Service (DNS), which allows resources (information, services and people) on the Internet to be easily located. It is the universal acceptance and wide availability of IP, DNS and a few other key protocols (such as HTTP for hypertext, SMTP for mail, ARP for address resolution, SNMP for management, TCP/UDP for packet delivery) that give the Internet its global reach, its broad availability and its vast user base.

One of the basic functions of the Internet is that it allows the transfer of information from one computer to another—and hence from one person to another. This was initially effected thanks to a file transfer application (known as FTP) which was designed to work with the TCP/IP package; many networked PCs have a pre-installed FTP client program. Simple FTP clients can connect to FTP hosts, view directories of files and download them according to user choice.

Typically, FTP servers require a username and password before they will allow connection. There are some FTP servers that will allow anonymous login; these require the username 'anonymous' and a password that is either your Internet Protocol (IP) address or your Internet email address.

The Internet Protocol

Internet addressing is based on the Internet Protocol (IP). This protocol, like any other, defines the expected behaviour between entities, a set of rules to which the communicating parties comply. IP is a network layer protocol that offers a connectionless service, that is, it does not establish a link between calling and called parties for the duration of a call. It is higher level protocols such as the Transmission Control Protocol (or TCP) that sit on top of IP that add this (and, with it, assurances of delivery).

Because of its frequent association with other protocols, the Internet Protocol is often referred to as TCP/IP. This is the collective term used for its many derivatives implemented on personal computers and UNIX workstations. In practise, Internet applications use either TCP or UDP (User Datagram Protocol): the differences between these two are that UDP is for unreliable connectionless packet delivery, and TCP is for reliable connection-oriented byte-stream delivery. UDP thus tends to be used for simpler services that can benefit from its speed, TCP for more demanding ones where it is worth trading a little efficiency for greater reliability.

Communications protocols can be (and in practice, always are) viewed as layered structures—a stack. This model has been used extensively to describe established telecommunications systems, and TCP/IP has not been exempt from the treatment, as illustrated in Figure 9.6.

The stacking or layering shown here is a convenient way to separate the various functions that work together to provide the familiar Internet services. Looking in a little more detail inside each of the layers:

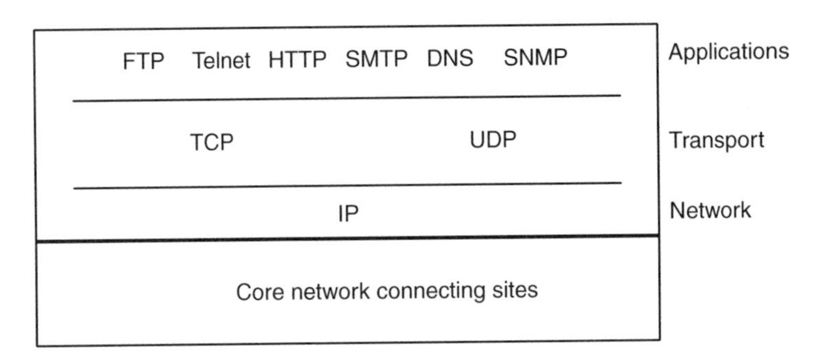

Figure 9.6 The four layers of the Internet

4 The *application layer* defines the application software, its processes and the protocol it uses to convey its data to the communications protocol stack. In the case of email, the protocol it uses is SMTP (Simple Mail Transfer Protocol). Email applications wrap up email messages with start and end markers and attach header information about who the mail is from and who the mail is to be sent to. This is passed to the layer below to be sent on its way; much like putting a letter in an envelope, writing an address on the front and dropping it into a postbox. Other protocols that reside in this layer are Domain Name Server (DNS), File Transfer Protocol (FTP), Simple Network Management Protocol (SNMP), HyperText Transfer Protocol (HTTP) and Telnet.

3 The *transport layer* wraps up the application layer message in its own data that defines the application is sending and the application to receive; these are known as the Source and Destination ports. It will also add data to specify the overall length of the message and a number, the checksum, to use to check if any of the data it is carrying has been corrupted. This is the layer in which both TCP and UDP reside. UDP is generally used to convey small messages of a request-response nature within single packets, and TCP is used to convey larger messages within a byte-stream.

The reasoning behind this difference is that, for small messages, the overhead of creating connections and ensuring reliable delivery is greater than the work of re-transmitting the entire message. To this end, TCP will attach further information to the message passed from the application to ensure that the reliable connection is maintained during the transmission, and that the segments of the byte-stream all arrive at their destination.

2 The *Internet layer* provides one of the most important functions of the TCP/IP stack by structuring the data into packets, known as *datagrams*. It also supports movement of the datagrams between the Network Access layer and the Transport layer, routing the datagrams from source to destination addresses and performing any necessary fragmentation and re-assembly of datagrams. The Internet layer wraps up the transport layer data in its own data which includes the length of each datagram and the source and destination addresses (the IP addresses) which specify the network and host of the source and destination.

1 The *Network Access layer* is perhaps the least discussed of all the layers, since the protocols within it are generally specific to a particular hardware technology for the delivery of data. Therefore, there are many protocols, one or more for each physical network implementation. The role of the Network Access layer is to ensure the correct transmission of IP datagrams across the physical medium and mapping of IP addresses to the physical addresses used by the network. The way in which the Address Resolution Protocol (ARP) is used to match hardware and logical addresses is explained it the main text.

The examples above explain how an application may send data across the Internet using the TCP/IP suite of protocols, from Application layer to Network Access layer. The reverse is also true when the Network Access layer receives data. The roles of each layer are the same, the difference being that the data is 'peeled' as it ascends the stack, each layer removing the layer-specific data put there by the source TCP/IP. The concept of wrapping up data, layer by layer, in this way is referred to as 'encapsulation'.

Addressing

From a network designer's point of view, one of the real beauties of the Internet lies in the way in that the IP datagrams are addressed to named hosts on the network.

Each host on the Internet has a unique number, known as an IP address (this concept, and the fact that there are a limited number of these, were discussed earlier in this chapter). The IP address is a 32 bit number that can be used to address a specific network and host attached to the Internet. This does not mean that every Internet user has a permanent IP address. For instance, dial-in users using the Point-to-Point Protocol (PPP) to connect to an ISP are 'loaned' one for the duration of their connection. Enduring IP addresses relate to a specific host on a specific network. The 32 bit IP address therefore needs to be split to define both a network part and a host part. This split occurs in different positions within the number, according to the class of the address, and there are four main classes in common use, A through D, as shown in Figure 9.7:

- Class A addresses are within the range 001 to 126.XXX.YYY.ZZZ—the first byte is used to define the network (bit 1 = class A, bits 2–8 the network). The remaining 24 bits are used to address the hosts on the network, therefore millions of hosts can be addressed.
- Class B addresses are within the range 128 to 191.XXX.YYY.ZZZ—the first two bytes are used to define the network (bits 1 & 2 = class B, bits 3–16 the network). The remaining 16 bits are used to address the hosts on the network, therefore thousands of hosts can be addressed from thousands of class B networks.
- Class C addresses are within the range 192 to 223.XXX.YYY.ZZZ—the first three bytes are used to define the network (bits 1, 2 & 3 = class C, bits 4–24 the network). The remaining eight bits are used to address the hosts on the network, therefore 254 hosts can be addressed from millions of class B networks.
- Class D addresses are within the range 223 to 255.XXX.YYY.ZZZ, and are special reserved addresses for multicast protocol address (they can be safely ignored).

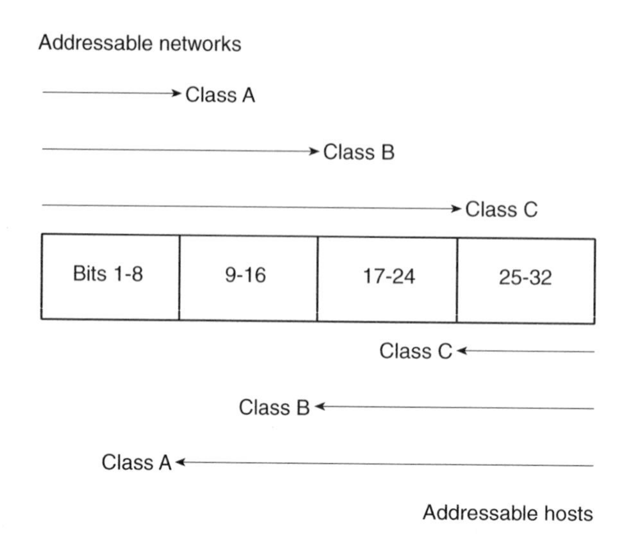

Figure 9.7 The 32 bit IP address

There are two class A addresses missing from the list above, these are 000 and 127. These are special addresses
that used for default and loopback routing—used when configuring a host. Also, host numbers 000 and 255 are reserved; 000 defines the network itself, and 255 is often used to broadcast to every host on the network.

Occasionally it is necessary to define additional networks from an address range. Using host address bits in the range as additional network address bits can do this. Thus, the total number of addressable hosts is reduced, but the number of networks increased. These networks are known as 'subnets'.

The need to define subnets is usually not technical but managerial or organisational. Subnetting caters for the delegation of address assignments to other organisations, however, subnets are only locally defined and the actual IP address of a host is still interpreted as a standard IP address.

Subnets are defined by applying a bit-mask (the subnet mask) to the IP address. If a bit is 1 in the mask it defines a network bit; if a bit is 0 in the mask it defines a host bit. It is best to define subnet masks on byte boundaries (it makes them easier to read); a standard class A address would therefore has a subnet mask of 255.0.0.0—all the bits in the first byte are 1—defining the network—all the bits in bytes 2, 3 and 4 are 0, defining the millions of hosts that are addressable on a class A network.

Before moving on, we should look at the familiar types of address that most people associate with the Internet. To do this, we take a Uniform Resource Locator (URL) and break it down into its constituent parts:

http://www.expl.com/expl.html

can be taken to bits, with:

- http://
  This part tells the client which protocol that it needs to use on this occasion. In this example, it is a standard access to a Web page, so HTTP is chosen. If the URL was targeted on a filestore, it's first part would most likely be ftp://.

- www.expl.com
  This is the readable version of the IP address as explained above. It resolves into the right address of the machine that you want to get your information from. So, assuming that the target machine is part of class B network, a familiar name for the intended server has to be resolved into something like 146.139.16.17—the server's IP address.

  The translation between familiar name and dot address is known as 'resolution', and is catered for by the Domain Name Service (DNS), which provides a hierarchy of lookup tables.

  Sometimes, there are qualifications to this address field (e.g. www.expl.com:8080). The part after the semicolon is not part of the address as such—it is there to route an incoming request to a particular port on the host machine. This may be included for security or operational reasons—the Secure Sockets Layer (SSL), for instance, redirects a web session to a different port.

- expl.html
  Finally, we get to the file that you are looking for. The '.html' extension should give you a warm feeling that all is well and that you are going to get an HTML coded document, as you would hope, using HTTP.

Many addresses are a lot longer than the one in our example, e.g., http://www.expl.com/net/faq/basics/expl.html. The intermediate references between the machine address and the final document reference (i.e. net/faq/basics) are simply there to get to the right part of the directory structure on the host.

Routing

Once all of the addressing has been sorted out, the next thing that needs to be done is to locate the intended recipient. This is where the Domain Name Service (DNS) comes in. It uses the dot address to see if the addressee is local. If not, it looks further afield, as illustrated in Figure 9.8.

DNS is central to finding any host on the Internet. Locating from among a vast number of machines is kept manageable by having a defined search hierarchy (Figure 9.9) that keeps the answer as local as possible.

Now that we know where we are going, the final step is to make sure that the network gets each packet to its intended recipient.

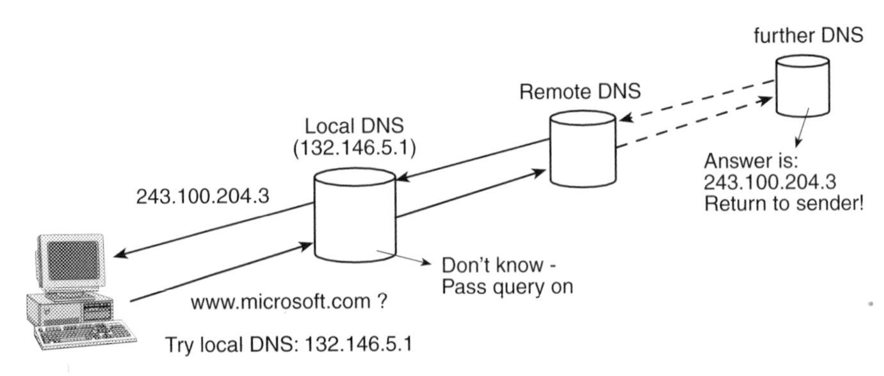

Figure 9.8 The strategy adopted by DNS for finding the addressed party

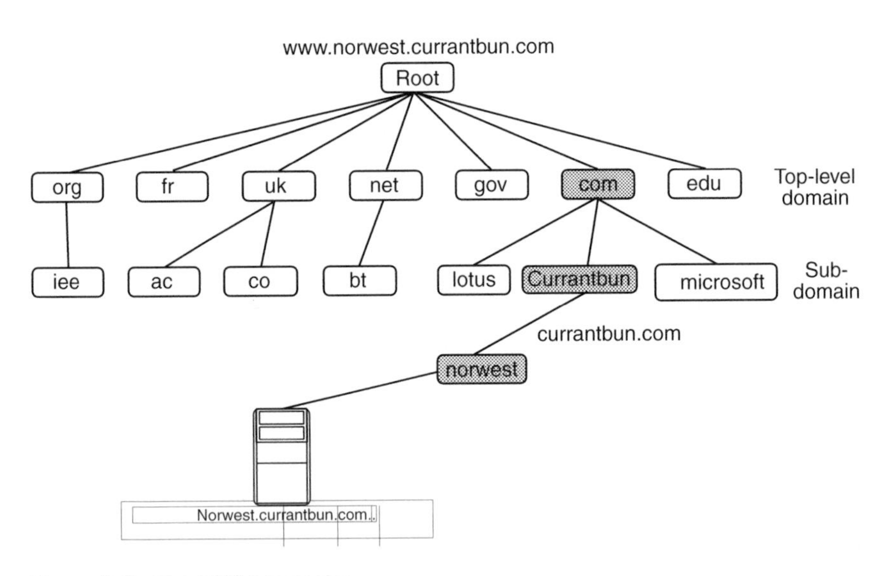

Figure 9.9 The DNS hierarchy

The Internet's routing strategy assumes that there are two basic types of device; hosts (a server or other computer) and gateways (e.g. a network router). Gateways are essential components of the Internet; without them, no two networks would be interconnected. Both these device types are required to route data both to and from hosts.

The routing decisions themselves are not complex. When a device, such as a router, receives an IP datagram it will examine the network portion of the IP address. It will first decide that if the destination address is on the local network it will deliver the datagram directly to the host. If it is destined for a remote network it will deliver the data to the next switch *en-route*, which will make the same decisions and route accordingly (Figure 9.10).

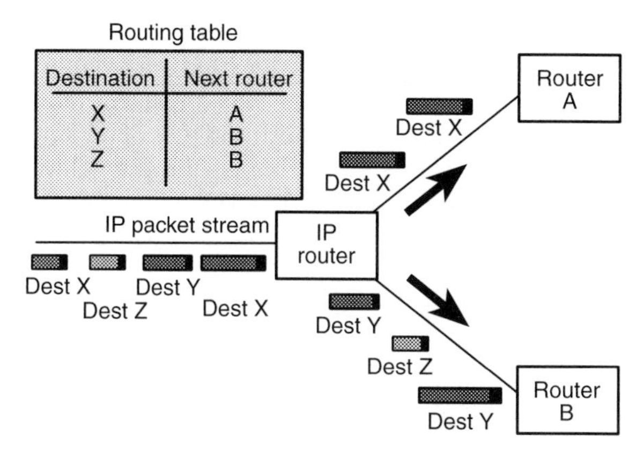

Figure 9.10 The routing decision taken at an Internet node

How the router decides where to go next depends upon the routing protocol being used. The most basic option is a Distance Vector Protocol, such as RIP (Routing Information Protocol). Routers running this type of protocol are just configured with a list of the subnetworks directly attached to that router. The router will periodically broadcast a routing update to all other routers attached to these subnetworks, telling them about the subnetworks that can be reached. When a router receives a routing update from a neighbour, the update will contain information about subnetworks not directly connected to the recipient router. The router will merge this newly acquired information into its own routing table. When this router next sends its own update, it will pass on information about the networks it has just learned about. The process is repeated until eventually, all nodes have learned routes to all subnetworks.

Although effective, RIP can generate a lot of traffic and the router can spend a lot of time updating its tables. To resolve these difficulties, a new generation of routing protocols has been created called the Link State Protocols; the TCP/IP version is known as Open Shortest Path First (OSPF). During normal working routers simply exchange 'hello' messages with their neighbours, to check that the links are working OK. These messages are short and use little bandwidth. Only if a hello handshake fails do the routers then enter a recalculation process, where all routers will recalculate the network topology.

When the network first starts up, routers will broadcast a message out of all their interfaces to try to discover its neighbours. Once a router has done this, it will broadcast a neighbour list to all routers in the network. The router is now able to calculate a complete network map from all the neighbour lists received. Only a failed link (or a new link added to the network) triggers a recalculation process. The knowledge of what is local and what is remote is usually pre-configured into gateways by system administrators. Gateways are also configured with routing information from more routing protocols. These protocols pass routing information between gateways to build routing tables that hold information on networks, and the extent of their interconnection. These routing tables, like those above, do not hold end-to-end routes; they hold lists of which networks can be reached via their own. There are a number of these gateway protocols: Gateway to Gateway Protocol (GGP), Exterior Gateway Protocol (EGP) Border Gateway Protocol (BGP) and Remote Gateway Protocol (RGP) are but a few.

More detail on all of the various elements of Internet technology can be found in a series of articles known as RFCs. This series may be titled Requests for Comment, but in practice it contains the definitive versions of all of the Internet's protocols.

9.4 DATA NETWORKS

The Internet uses distributed intelligence to take advantage of existing networks, with each connected device contributing to the overall power of the network. This contrasts with the familiar telephone network, where a fixed hierarchy of switches exercises control. The Internet also differs in the way in which 'calls' are made. The traditional telephone call is made by setting up a path from one user to another. That end-to-end path is dedicated for the duration of the call, an arrangement called 'circuit switching'.

An Internet call uses an alternative approach, called 'packet switching'. This entails splitting a message into small pieces (packets) which (like mail) are individually labelled to indicate their destination address. The system then uses whatever resources are available to deliver the packets (and hence the complete message). It is an efficient and flexible way of doing things, but it is very reliant on the underlying connection. The most common of these is the plain, ordinary telephone service—a phone line and modem is what most people use to get online. There are other options, though, ranging from the ISDN (which can be thought of as a go-faster version of the ordinary telephony network) through to high-speed, special purpose data networks like SMDS. The main difference between them, at least from a user point of view, is speed and cost. There are also significant differences in the way that each type of network operates, and this can have quite an impact of its suitability. So we now look in some detail at each of the main technologies used to build commercial data networks.

ISDN

ISDN stands for Integrated Services Digital Network. As the name suggests, it is an all digital network that allows a range of different services to be carried together on the same circuits. It can be considered to be an extension of the public switched telephone network, the key similarity being that it permits any two compatible pieces of connected equipment to talk to each other. This means that ISDN can carry any form of data, such as voice, video and computer files.

The motivation behind ISDN was to replace the analogue telephone network with a less noisy, digital one. It was designed around the notion of separate channels operating at 64 k-bits per second (a number that springs from the fact that basic, analogue voice transmission requires 8k samples per second, each of which is encoded as 8 bits). Although no more than satisfactory for voice, 64 kbits/sec is between two and five times the average speed that most computer users enjoy. So, one of the main reasons for many people to be interested in ISDN is that it provides a significant improvement for online applications. There is a noticeable improvement in quality with video and desktop conferencing, World-Wide Web access and the transfer of large data files

In the UK and Europe, ISDN is offered in two forms, ISDN2 and ISDN30, where the number suffix denotes the number of 64k channels that are provided. ISDN2, also known as Basic Rate Access, gives you two 64k (B or bearer) channels and a single 16k signalling (D or delta) channel. ISDN30 is also called Primary Rate Access, and provides 30 B channels along with a D channel. In the US, Primary Rate Access is based around 24 B channels, with one D channel. In both cases, Basic Rate is intended for home use, and Primary Rate is meant for businesses.

In practice, there are many instances where you might want some number of channels between 2 and 30. High quality video-conferencing, for example, requires the capacity of about six B channels. There are several approaches to getting the right speed to suit a wide variety of services, using techniques known as inverse multiplexing. The most common method (called BonDing after the Bandwidth on Demand Interoperability Group) can be used along with standard ISDN channels to support up to 63 combined B channels. Other options include Multilink PPP (designed for Internet traffic over ISDN) and Multirate Service (an $N \times 64$ service, provided as part of the ISDN service).

By way of contrast, there are times when you want to use a 64k channel to carry data of a lower speed. This is known as 'rate adaption' and, again, there are standards for this. The two most common are known as V.110 and V.120. Both ensure that slower data is carried safely over the higher speed bearer.

To become part of the Integrated Services Digital Network, you first have to be connected to a digital exchange and be within the ISDN transmission limit. In the UK virtually all exchanges are digital, and over 90% of customers can be successfully hooked up.

The ISDN equivalent of the telephone socket, at least in the UK, is called the Network Termination Unit, or NT1. This is a box that has copper wires going back to the main telephone network on one side and a socket, like the standard phone socket only a bit wider, on the other. If you have ISDN compatible equipment, you can plug it directly into the NT1. If not, you need a terminal adapter or TA. This is used to connect ISDN channels to the interfaces you get on most current computing and communications equipment (e.g. the familiar 25 pin RS-232 connector).

The precise details of where and how you connect varies from place to place. ISDN standards use a set of reference points (such as the S/T interface between the NT1 and the TA) as a basis for interworking between devices, and to define the boundary between the phone network and your private installation. This is shown in Figure 9.11.

With two main channels and a separate signalling path at your disposal, there are many options for using ISDN. For instance, you can hook up to eight devices to one ISDN line, and these can be placed anywhere on a bus connected to the S/T point. There are limits on how far this bus stretches (about 200 metres), and you have to provide power for all devices, but it is conceivable that you could control your microwave, turn on the house lights, send Internet mail and phone a friend all at the same time!

Some of the equipment that allows you to exploit the ISDN is readily available. You can buy ISDN adapter cards that plug into your computer, just like a modem. More sophisticated devices, such as those that allow you to build your own home network at the S/T point, is also becoming available. Overall, ISDN gives the domestic Internet user a faster, more flexible means of connection.

xDSL

Just as ISDN extended the usefulness of the local line (i.e. the private customer's link to a public telecomms network), so the xDSL technologies

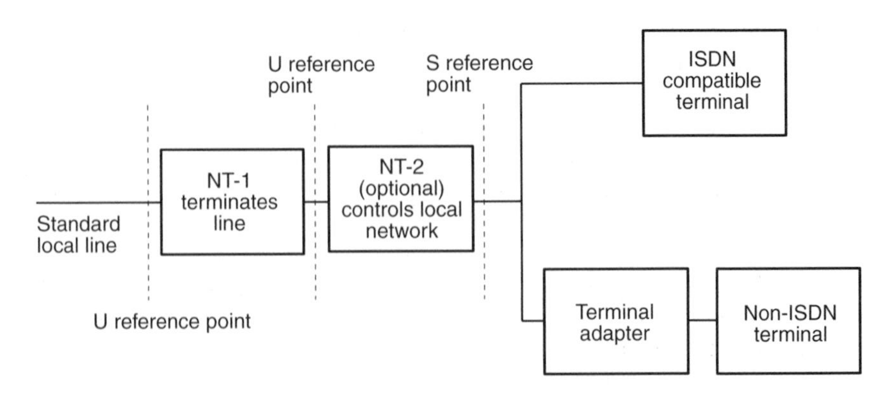

Figure 9.11 The logical arrangement of an ISDN local end

promises to do the same thing. It allows local access in the Mbit/sec range for existing phone users, albeit with some restriction imposed by the length of the line. xDSL itself defines how a pair of modems—one located at the local telephone exchange and the other at the customer site—can be used to deliver high-speed signals over their established twisted-pair copper connection. A typical installation is shown in Figure 9.12.

There are several varieties of xDSL:

- Asymmetrical DSL (ADSL) allocates the available bandwidth in an asymmetric spectrum so that core data is delivered downstream (toward the user), then is returned to the exchange in the upstream channel. ADSL is well suited for high-speed Internet/Intranet access, video-on-demand and telecommuter applications. ADSL speeds range from T1 (1.544 Mbits/sec) and E1 (2.048 Mbits/sec) to 6 Mbit/sec and beyond downstream. Upstream return channel speeds range from 64 kbits/sec to 384–640 kbits/sec. ADSL transmissions operate at distances up to

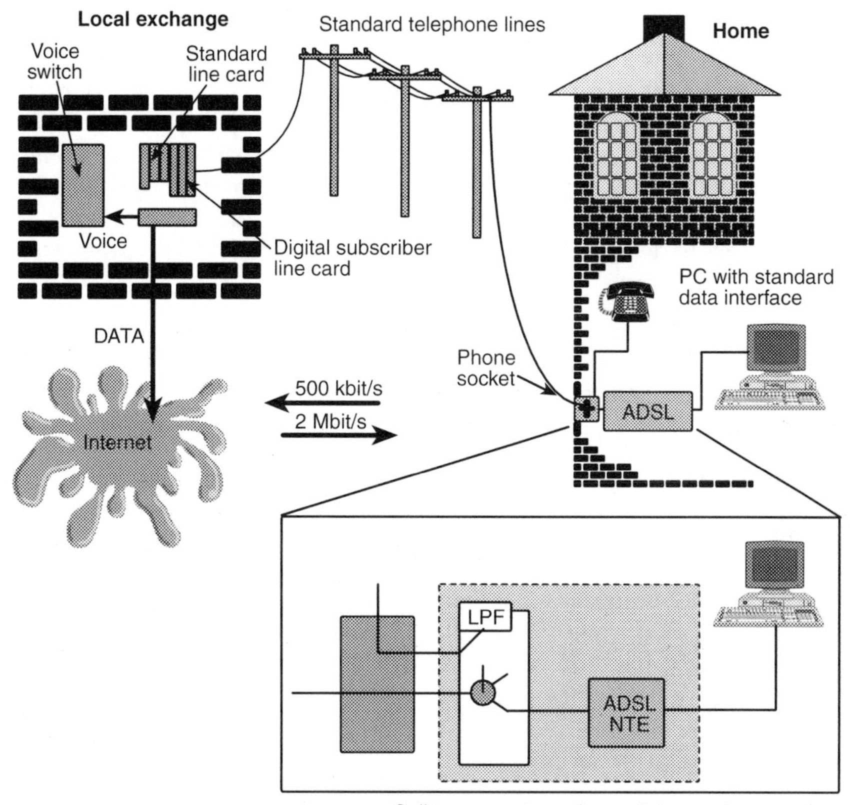

Figure 9.12 Deployment of ADSL over a local loop

5km (between the customer and the local exchange or serving central office switching system) via a single copper twisted-pair.

- High-Speed DSL (HDSL) is a symmetric technology with speeds of 1.5 or 2.0 mbits/sec (upstream and downstream). Its main purpose is to replace traditional T1/E1 leased circuits. As for the standards bodies, HDSL is a two-wire implementation with an operating range somewhat more limited than that of ADSL—much over 3km and telephone companies need to install signal repeaters to extend the service. Because HDSL is a two-wire implementation, it is deployed primarily for PBX network connections, digital-loop carrier systems, interexchange Points of Presence (POPs), Internet servers and private data networks.
- Symmetric DSL (SDSL) is similar to HDSL in that it delivers 1.5M or 2.0 Mbits/sec (or submultiples), but it does so using a single line, downstream toward the user and upstream. The use of a single line further limits SDSL's operating range; 10,000 feet is the practical limit for SDSL applications. Because of its symmetrical nature, it is well-suited for videoconferencing applications or remote LAN access.
- Very High-Speed DSL (VDSL) is asymmetric. Its operating range is limited from 1000 to 4500 feet, but supports very fast transmission via single twisted-pair copper. Data can travel at rates up to 51.84 Mbits/sec from 330 to 1000 feet with rates of up to 1.6 Mbits/sec on the upstream return path. VDSL is positioned as the eventual modem of choice for fibre-based, full-service networks. The extra bandwidth allows telephone companies to deliver High-Definition Television (HDTV) programming using VDSL technology.
- Rate Adaptive DSL (RADSL), which automatically adjusts to copper quality degradation or can be manually adjusted to run at different speeds up to ADSL rates.
- IDSL or ISDN-based DSL, which 'inverse multiplexes' two ISDN 64 bits/sec B channels using 2B1Q coding into one 128 kbit/sec channel.

All of the DSL technologies have been subject to some work within ANSI, the body that sets US standards for telephone line transmission, and ETSI (European Telecommunications Standards Institute), its European counterpart. The ITU-T (formerly CCITT), the worldwide telecommunications standards body has not yet addressed DSL, because the systems are intended for local network services and so do not generally cross national boundaries. Despite this, most countries are in the process of introducing a commercial ADSL service.

Frame relay

Frame relay is very much a product of the digital age, exploiting the much lower error rates and higher transmission speeds of modern digital systems. In particular, frame relay has its roots in the Integrated Services Digital Network. As a development of the PSTN, the ISDN is intrinsically circuit-switched, but packet-switching is more relevant for the Internet user. The first generation of packet switches (based on the X.25 protocol) do not fit the ISDN model of keeping user information and signalling separate and use heavyweight error correction protocols in the comparatively error-free digital environment. Something else is clearly needed in ISDN to support data services effectively, and frame relay was defined to fill this gap.

Frame relay is a simple virtual circuit packet service developed as an ISDN bearer service for use at data rates up to 2Mbits/s. It provides both Switched Virtual Calls (SVCs) and Permanent Virtual Circuits (PVCs), and follows the ISDN principle of keeping user data and signalling separate from each other (cf. a normal telephone call, where the signalling and speech share the same bandwidth).

An ISDN Frame Relay SVC would be set up in exactly the same way as an ordinary circuit-mode connection using ISDN common-channel signalling protocols. The difference is that in the data transfer (or conversation) phase, the user's information is switched through simple packet switches—known as frame relays—rather than circuit-mode cross-points. The network operator can set up PVCs on subscription: user signalling is neither needed nor provided.

In frame relay information is transferred in variable length frames. In addition to the user's information there is a header and trailer. The header contains a ten-bit label agreed between the terminal and the network at call set-up time (or at subscription time if a PVC) which uniquely identifies the virtual call.

Terminals can therefore support many simultaneous virtual calls to different destinations, or even a mixture of SVCs and PVCs, using the identification facility to identify to which virtual call each frame belongs. Values from 16 to 991 are available to identify the user's SVCs and PVCs (others are reserved for specific purposes—0 is used to carry call-control signalling, 992 to 1007 carry management information).

One of the great merits of the simple data transfer protocol is that it provides a high degree of transparency to the higher layer protocols that are carried. This contrasts with the older X.25 protocol, where the scope for interaction with higher layer protocols often causes problems, and can seriously impair performance and throughput.

However, simplicity has its price. The absence of flow control leaves the network open to congestion; congestion ultimately means throwing frames away. Throwing frames away causes higher layer protocols to re-transmit lost frames, which further feeds the congestion, leading to the possible collapse of the network. Congestion management is therefore an important issue for the network designer and network operator if these serious congestion effects are to be controlled and, preferably, avoided.

Given the flexibility of frame relay, it is important for the users and the network operator to agree on the nature and quality of the service to be provided. This gives the service provider an estimate of the traffic to be expected, essential for properly dimensioning the network, and it gives users defined levels of service which they can select to best match their requirements. The frame relay standards specify a number of parameters that characterise service quality, some relating to the demand the user will place on the network, others specifying the performance targets the network operator is expected to meet.

One of the dominant trends of the last decade has been the rise and rise of the personal computer. From its early use as a word processor, it has evolved to become an indispensable part of a company's information infrastructure, and is now almost as common in the office as the telephone. One of the most remarkable features of this evolution has been the associated almost explosive growth in Local Area Networks (LANs) used to interconnect them.

By their very nature, LANs are capable of only limited geographical coverage. Companies with LANs in different locations therefore need to interwork them, often over long distances and sometimes internationally. A confusion of bridges, routers, brouters, hubs, gateways and other devices have been developed to adapt the profusion of proprietary LAN protocols to the available wide area channels used to interconnect them.

Until frame relay came along, the choice for wide-area LAN-interconnection lay between leased lines and X.25. The former tended to be expensive, especially for international interconnection (and, in truth, still is), and not well matched to the bursty nature of LAN traffic. The latter is a complex protocol which tends to interfere destructively with any higher layer protocols being carried, usually degrading throughput seriously, often severely, and occasionally fatally.

Frame relay's high speed and transparency to higher layer protocols make it an almost ideal choice for interconnecting LANs over wide areas. Hence, frame relay provides an ideal service for many Internet users, one that is well matched to the type of traffic being carried in small to medium businesses.

SMDS

SMDS (Switched Multi-megabit Data Service) is a public, high speed packet data service that, like frame relay, is aimed at wide area to LAN interconnection. The most commonly difference cited between the two is that SMDS is considerably faster—up to 34 Mbits/sec is available. In fact, the two are fundamentally different in the way that they operate. This may or may not be of significance to the end user, but it is worth spending a little time on the principles of SMDS.

The first thing to appreciate is that SMDS is a connectionless service. This means that every packet contains full address information to identify both the sender and the intended recipient(s). SMDS also enforces a point of entry policy such that the network access path can be dedicated to a single customer. Together, these mean that addresses can be screened and communication restricted to selected users. So, even though it is based on a public network, SMDS can offer the security of a private network.

Four key service characteristics are defined within SMDS. These are a connectionless service with multicast capability, global addressing, a facility for creating virtual private networks, and a mechanism that protects users from paying for unused capacity. In terms of access speed, the standard classes used are at 4, 10, 16, 25 and 34 Mbits/sec. Higher rates of 140 and 155 Mbits/sec are planned for the future. With the current range, 10 Mbits/sec Ethernet LANs or 16 Mbits/sec token ring LANs can readily be linked.

A typical SMDS implementation would have a router accepting IP packets that would be attached to an SMDS packet and sent over the SMDS network, where the reverse process would take place. The way in which the IP packet (or any other packet, for that matter) is connected to the SMDS packet is called the SMDS Interface Protocol (SIP). A variety of technologies can operate over SMDS, thanks to the fact that a clear interface is defined. A data exchange interface (DXI) and a relay interface enable a wide range of end devices to be connected via SMDS.

Perhaps the best example of SMDS is its application in SuperJANET, the high speed academic network in the UK. It provides the backbone that connects around 70 universities, and is used for a wide variety of applications, including the transmission of medical images, video-conferencing and education material.

Since SMDS and frame relay fulfil much the same function (interconnection of LANs), the choice between them will usually be in terms of price and performance, and this is specific to user need and operator tariffs.

ATM

Asynchronous Transfer Mode (ATM) is seen by many in the telecommunications industry as the basis for a true multiservice network. Unlike SMDS and frame relay, ATM is a technology rather than a service. It provides a high speed, connection-oriented mechanism for carrying both, so is not an alternative to SMDS or frame relay but is a support to either.

The basic operation of ATM is to route short packets of uniform length (called cells) at very high rates. Each cell is 53 octets (eight bits) long—up to 48 octets of data and five octets of addressing and control information. This uniformity of structure allows the switching of cells to be carried out by static hardware rather than software, and it is this feature that underpins the high operational speed of ATM. In action, ATM works by waiting until a cell payload of user information is ready and then adding a cell header before passing the complete cell on to a local switch, where the cell is routed through the network to its destination. No regard is paid to the content of the cell in this mechanistic part of the process—a uniform switching is presented to all types of traffic, hence the suitability of ATM to multiservice networks. A variety of control bits are included in the header to secure effective delivery.

The fact that ATM can carry a whole range of different types of traffic—voice, video, text—cannot be ignored. Because each of these types of data has different requirements in terms of delay or error tolerance, there are a number of options defined within ATM for putting the raw information into the cells. Within the descriptive model of ATM, there is a layer known as the adaptation layer that copes with this. The ATM Adaptation Layer (AAL) sits between any service-specific functions and the basic cell assembly layer. There are four distinct AALs, each defined to support a different class of service (e.g. connection-oriented with constant bit rate, connectionless with variable bit rate). The features provided within each AAL come at a cost of reduced payload—the more sophisticated the facilities required by a service, the more payload is used to provide it. Over and above these coding options, there are also defined classes of service that allow paths and circuits to be managed effectively.

It was mentioned above that ATM is connection-oriented. Creating suitable entries in lookup tables in every switch *en route* makes these connections. There are two options for doing this. If the entries are made at subscription time, a Permanent Virtual Circuit (PVC) is created. If, on the other hand, the entries are made at call set-up time, a Switched Virtual Circuit (SVC) is created. The latter places greater demands on switch design and has been a less common option.

There is a lot of flexibility built into the way that ATM connections are established. The connections established between two sites, for instance, can be further divided into a number of virtual paths. This allows flexible interconnection of user sites—for instance, a connection may support a link between private exchanges, a videoconferencing link and a frame relay service. ATM cards that enable a PC to connect via a local area network have been available since the mid 1990s, and ATM network infrastructure is developing quickly, both in the UK and in other countries. A considerable amount of detail on these network technologies can be found in the companion books in this series, *Total Area Networking—ATM,IP, Frame Relay and SMDS Explained*.

9.5 MOBILITY

The requirement of fast data access for mobile users has spurred technology advances in wireless networks. GSM currently provides up to 9.6 kbit/sec, but the next generation will increase this to 64 kbit/sec. General Packet Radio Service (GPRS) will enable IP services to be delivered over a GSM-type packet switched network.

Future mobile networks, such as the Universal Mobile Telecommunication System (UMTS), aim to increase data rates to 1–2 Mbits/sec. UMTS may become a broadly used network, offering both higher bandwidth and additional services, such as:

- Terminal mobility, by which it should be possible for devices to access services from different locations while in motion, the network locating and identifying the device and user.
- Personal mobility, enabling a user to access services at any fixed or mobile terminal on the basis of a personal telecommunication identifier.
- Service mobility, providing access to the same set of services from any connection point within public and private environments. This concept is also known as the Virtual Home Environment (VHE).

Mobile operating systems

One of the points made in the section on access devices was that each tends to have its own operating system. The current view is that no single Operating System (OS) will dominate the mobile computer market, but there are four main contenders.

The Palm OS, owned by 3Com, is the basis for the successful Palm palmtop device range. At the end of 1998 it had the largest PDA user base, with over 40% of the worldwide handheld market, and 70% in the US.

Microsoft released Windows CE a few years ago with the vision that it would suit a range of small devices, including handheld PCs, palmtop computers and auto PCs (e.g. car stereos). It has failed to make the market impact expected on its initial release, but with many established hardware vendors (including Philips, Casio, Hewlett Packard and Sharp) increasing their device range, and enhanced OS releases, it may see significant market growth.

The third vendor is Symbian with its EPOC operating system. The Symbian alliance includes Psion, Motorola, Ericsson and Nokia, producing an interesting combination of PDA, CPU and mobile phone technology manufacturers. EPOC is used in the current Psion PDAs, but no mobile devices using EPOC have been announced yet.

Java OS is the fourth option, and Sun have released a specification detailing a Java Application Environment, PersonalJava, which is aimed at consumer devices. Its architecture is more suited to constrained memory and processing environments, and can optionally include parts of the full Java Application Programming Interface (API). New devices running Personal Java operating systems are only just appearing, and it is difficult to predict their market take-up. Sun are also producing other Java APIs for speech, sound, telephony and multimedia. These are likely to migrate into mobile devices.

The other consideration with Java is that Java Virtual Machines (JVM) may be available on top of existing OSs. Symbian have released details of their intended implementation of a PersonalJava JVM, to run on the EPOC OS, and a JVM for Windows CE OS is available.

Web access for mobiles

Most existing e-commerce services use Web access, and as this is likely to continue, mobile computer Web browsing capabilities will be important. Laptop and notebook computers use standard Windows PC Web browsers, although performance over a slow wireless network can be frustrating for pages with high graphics content. Web browsers for smaller devices, such as personal organisers, are quite varied and often tightly linked to the device platform. Comparing the four main vendors in the area:

- 3Com rely on independent software vendors to supply Web browsing capabilities for their Palm devices, including Smartcode's HandWeb, which offers a simple text only interface, and ProxiNet's ProxiWeb browser, which performs some proxy image conversion.
- Microsoft Windows CE offer a cut-down version of the Desktop PC Internet Explorer.
- Symbian offer their own Web browser for EPOC OS.
- Java OS has Sun's own HotJava browser, although they will be offering a package of Personal Applications to suit PersonalJava environments.

For wireless devices, the Wireless Access Protocol (WAP) group is developing standards for Web access. WAP compliant devices use a specialised Wireless Mark-up Language (WML), with specially developed content or via a proxy translation from HTML to WML. They use a menu driven user interface (micro-browser) to access Web text data.

Wireless Knowledge, a Microsoft and Qualcomm joint venture, has proposed a proprietary solution. All wireless enabled devices connect to a central operations server, which provides interfaces to different networks (particularly attractive in the US with a multiplicity of network operators and technologies) and device types. Qualcomm have agreed to work towards engineering Windows CE into their wireless handsets, to be sold under the Qualcomm brand in the US and other brand names in foreign markets.

In the mobile arena, web access options are becoming very diverse. If an e-commerce service is dependent on certain features, then either users need the correct Web browsing facility or the service has to adapt itself to the specific characteristics of the access device. This is the issue that adds significant complexity to e-commerce service delivery. Even though web access may soon be available from many devices, it is essential that the eBusiness service remains presentable on the access device being used, either through adapting the content at the service provider end or within the device.

9.6 ELECTRONIC CASH STANDARDS

The concept of using a smart card as an electronic purse is not new, but it is only just beginning to take off. The idea is simple enough: funds are transferred from a bank account to the card at a terminal and are subsequently transferred from the card to a merchant when used to pay for a variety of products and services. It is a pre-payment scheme, much like telephone cards of today, but greatly superior since payment by electronic purse can be accepted by an unlimited number of merchants unlike the 'closed' telephone card application.

Many of the benefits of electronic purse technology lie at the merchant/ bank side as less cash has to be handled, the risk of robbery is reduced and the payment process is faster. Customers reap some reward from potentially faster payment at tills and vending machines (although slow speed was actually a problem at recent trials of Mondex and VisaCash), and do not have to carry the correct change all the time. There are some interesting downsides for customers, though, such as what might happen if they lose their card, or the purse operator goes into liquidation. In addition, they are essentially granting the operator an interest-free loan, since they exchange real money for electronic money that they may not spend for a period of time.

Proton

The Proton scheme is the most successful international electronic purse standard; 30 million such cards having been issued in 15 countries worldwide. A number of national schemes, such as ChipKnip in the Netherlands, use the standard. American Express and Visa both have licences to use it. The Proton specifications are now administered by a profit-making company called Proton World International, whose members include Visa, American Express, ERG, Banksys and Interpay.

VisaCash

VisaCash has issued around 21 million cards, although this figure is misleading as it includes their low-security disposable version, which cannot be classed as an international purse technology. It is Visa's policy to badge existing national purse schemes where possible, so the brand includes a variety of purse standards. A large number of trials and pilots have been running in various countries for several years.

A joint trial of VisaCash and Mondex is currently running in Manhattan, US, but it is perceived by consumers and retailers as not worth the effort, with slow transaction speeds said to be the biggest problem.

Mondex

Mondex, whilst being the most technologically advanced, is the smallest by far of the internationally accepted schemes, having issued one million cards. In trials, usage of the payphone option was quite low, but the home phone was relatively popular. Trials are also running in Canada and Australia, and there has been a full roll-out in Hong Kong. The latter is proving to be a commercial success, being popular with both consumers and retailers. Mondex is the only purse scheme that emulates real cash, since transactions are not centrally accounted. This property disintermediates some of the stakeholders, and makes it difficult for governments to exert fiscal control over the e-cash economy. This has so far prevented Mondex from penetrating the European market to any great degree, outside the UK.

Visa, Proton and a number of European financial institutions are collaborating to create the Common Electronic Purse Specification (CEPS). This development is spurred on partly by the advent of the new common European currency, the Euro. It is expected that all national purse schemes wishing to have some degree of international inter-operability will migrate to CEPS. The Euro/CEPS combination could be the biggest driver for the adoption of electronic cash.

Various small and medium-scale trials have been run of all the above purse schemes, to show the viability of electronic cash for making small value e-commerce payments. The technology works well, especially when combined with user-friendly payment wallets, integrated with web browsers. The only barriers are the lack of interoperability standards for electronic cash and the lack of infrastructure of smart card readers on e-commerce access devices. The removal of these barriers is imminent, with the arrival of PC/SC and Smart Cards for Windows, as discussed earlier in this chapter. These barriers do not apply to the use of e-purses with GSM equipment, and Cellnet is currently gearing up for a trial of loading value onto Mondex cards over GSM, i.e. a mobile ATM.

It is widely accepted that there is no short-term business case for electronic cash in high street transactions. Deployment, therefore, has to be looked at from a perspective of long-term protection of one's business. Financial card issuers will therefore slowly deploy electronic purse on the back of other applications, the marketing model for which does not seem to be understood. Electronic cash will add value to the e-commerce world, but since e-commerce currently represents only a tiny fraction of the world's cash economy, benefits for institutions are currently small. Again, progress is slow, but could then explode as standardised purse schemes and interfaces to PCs come onto the market.

9.7 MARKUP LANGUAGES

We are all familiar with WYSIWYG word processors which allow the user to produce documents that employ the sort of complex formatting that was once the province of the professional publishing industry. The most humble word processor probably allows the use of different fonts, bold and underlined text, distinctive titles, indented text and a whole range of other features for customising the appearance of documents. It follows that when a word processed document is saved onto a computer disk, not only is all the text, or 'content' stored, but also a mass of information relating to the layout and appearance of the document must be saved too. Each word processor manufacturer has its own proprietary method for encoding all this structuring information that must accompany the basic text. It is the difference between these methods that leads to the incompatibilities that often prevent a document created using one word processing package to be viewed using a different manufacturer's product. In fact, these word processor file formats are so intimately bound up with the particular formatting features offered by a particular product that there is often an incompatibility between document formats generated by different variants of the same vendor's product.

Word processors use a variety of ways of encoding formatting information. Some embed cryptic sequences of codes interspersed in the text to indicate such features as bold text or changes of font. Others embed pointers within the text which refer out to a table of 'styles' which are defined (again in an encoded format) at either the beginning or end of the document file. All these share a number of features:

- They are not 'human readable'. They can be interpreted by the word processing application, but not by a person who tries to decode the file.
- They are not 'standards' (either amongst vendors or even product versions of the same vendor). Neither are they generally published or open.
- They are designed to optimise the editing process from the end user viewpoint. This means that they are optimised for efficient handling by the computer.

Whilst word processors have been developing with ever more powerful features and proprietary file formats have evolved with corresponding complexity, a separate strand of activity has been underway in the industry, to develop 'markup languages'. Like a word processor file, a markup language is intended to encode information about a document's structure and content. However, the concept of markup has its origins in the publishing industry, with the marks that authors or editors employ to indicate how parts of the text should be formatted in the typesetting and printing stage of a document.

Hence, computer markup languages are human-readable (if a little cryptic) and are made up of characters accessible from the normal computer keyboard. So that they can be distinguished from the textual part of the document, they have to have some distinguishing character sequence that instantly identifies them as markup sequences. For example, the most commonly used markup languages which we shall be discussing, enclose their markup 'tags' within a pair of angle brackets: < >. For example, the symbol for the start of a new paragraph could be <P>; the symbol for a title, <TITLE>, and so on. There have been considerable attempts to standardise markup languages internationally.

To further clarify the difference between a markup language and a word processor file, we can think of a document as containing three things:

- The content: the words, pictures and other items that it contains. In a multimedia document this can include sounds, video or any other media.
- The structure of the document. How it is divided into paragraphs, headings, chapters or similar items.
- The rendition: the way that the structure is portrayed on the screen or printed page. What size and weight of font is to be used for titles; how the break between chapters is to indicated; and so on.

In the case of a word processor, the top feature is generally usability, and that means that the 'What You See Is What You Get' feature is paramount.

This in turn leads to the three things being mixed up together. On the other hand, markup languages seek to separate these out, and are aimed primarily at structure.

In particular, markup languages do not generally make any statements about rendition. This may seem at first sight to be a weakness: what is the use of a language that tells you the structure of a document but doesn't tell you how it should appear on the page? In fact, this becomes an advantage when one is dealing with multimedia documents delivered over a medium such as the Internet. One of the problems of such a delivery is that the author and publisher have no knowledge of the particular computer that will be used by their reader. Hence, there is advantage in delivering a document in the form of content plus structure and leaving it to the client computer to decide how to render it on the screen.

We now discuss the main developments in markup languages, and their application in presenting electronic goods.

SGML

Standardised Generic Markup Language (SGML) is the most important markup language that has been standardised internationally. It is sometimes described as a 'meta language', meaning that rather than being a complete markup language in itself, it is a actually a language framework that can be used to define any number of markup languages for specific purposes. As we shall see, one such use of SGML has been to define the hypertext markup language HTML, which has been central to the development of the World-Wide Web and hypertext publishing.

SGML, in keeping with the philosophy of markup languages that we have discussed, sets out to encode (using markup tags) the structure of a document, and not its rendition. For example, we might decide that there is a particular type of document called a novel, whose structure consists of a book title followed by a series of chapters. Each chapter begins with a chapter title followed by a series of paragraphs. To describe the structure of such a document, we might define the following markup tags:

Start of novel: <NOVEL>
End of novel: </NOVEL>
Start of book title: <BTITLE>
End of book title: </BTITLE>
Start of chapter: <CHAPTER>
End of chapter: </CHAPTER>
Start of chapter title: <CTITLE>
End of chapter title: </CTITLE>
Start of paragraph: <P>
End of paragraph: </P>

There are several things to note about the markup tags: the first is that they are always enclosed in angle brackets, to distinguish them from the content of the document; the second is that the end tag for a particular element of structure is indicated by the solidus '/' sign.

If a simple novel was structured using the above markup, it might look like:

```
<NOVEL>
  <BTITLE>
  The Gumbleton Saga
  </BTITLE>
  <CHAPTER>
    <CTITLE>
    In the Beginning
    </CTITLE>
    <P>
    Josiah Gumbleton never forgot the first day he arrived in London. He
had taken the coach from Deal and had arrived at the Turk's Head in
Newgate Street. From there he had hired a cart to take himself and his wife
to the their new premises in East Lane, Walworth . . . </P>
</CHAPTER>
<CHAPTER>
  <CTITLE>
    The next instalment
  </CTITLE>
    The next year . . .
</CHAPTER>
</NOVEL>
```

Various features are apparent. In general, elements of the document are enclosed in matching sets of tags such as <CHAPTER> . . . </CHAPTER>. Some tags are not necessary; their effect can be inferred from the context. For example, the end of a paragraph is always followed by the start of a new paragraph (a <P> tag) or the end of a chapter (a </CHAPTER> tag). Hence, the </P> tag is not strictly needed to mark the end of a paragraph. Some tags might appear anywhere (<P> paragraph tags could appear almost anywhere) whilst others can only appear in certain places (<CTITLE>, for instance, can only occur within a chapter).

SGML provides a language for defining tags, and includes such characteristics as we have just encountered:

- whether it must have an obligatory matching end tag;
- whether it can appear anywhere or only nested within some other pair of tags.

A set of tags defined using SGML for a particular purpose is known as a Document Type Definition (DTD). The markup languages widely used for Internet-based multimedia are defined as SGML Document Type definitions.

HTML

Hypertext Markup Language (HTML) is the most widely used DTD for online multimedia systems. Although HTML is formally defined in terms of an SGML DTD, most practitioners make full use of HTML without delving into the complexities of SGML.

HTML is used to define the layout of a computer screen combining text (in a range of typographical styles) and images, whilst combining the ability to address other forms of information through hypertext links.

At the heart of hypertext is the anchor. This is a very simple construct within the HTML specification that identifies text, a single word or a string, as a hypertext link. An anchor is typically of the form:

Just an <A HREF="http://www.expl.com/expl.html">example</A> of an anchor.

Where the link, 'example', would be shown by the browser in blue or some other colour rather than the default black. To click on the anchor would result in a request being sent from the client to the server (http://www.expl.com/) defined in the Hypertext reference (HREF). The request would be sent as an HTTP request, asking for the file 'expl.html'. The file would then be sent to the browser and displayed.

If a file is not an HTML document, it will usually have a suffix that denotes its type (e.g., .txt, .doc, .ppt, etc.). When the server encounters this suffix it will prepare the file according to its local type definition within its configuration, and send it to the browser. The browser, on encountering a file type that is not HTML, will also look-up the file suffix in its helper application configuration and, if it is defined as suffix type or associated with a particular application, it will either save the file to disk or boot-up the application so that it (and not the browser) can handle the file. For example, a file with suffix '.pdf'[1] will result in the browser booting Adobe Acrobat for it to be read.

HTML is easy to read and write, since it consists of plain text delimited by tags (the special characters in chevrons), and for this reason, it is easy to

[1] pdf stands for Portable Document Format. It is a standard for document presentation that helps to ensure that a page of information, when viewed, is in the format that the originator intended.

see just how extensible this language is. For example, an anchor always starts with <A . . . > and ends in </A>.

To give some idea of its pace of development, HTML has seen at least two revisions a year since it went under change control, and now includes a whole raft of features, from tables to text fields, buttons, check boxes and context-sensitive image definitions.

eXtensible Markup Language (XML)

SGML and HTML both focus on how online information should be presented. XML goes a little further, in that it aims to define what type of content is enclosed in a document. As such it is an ideal candidate for enabling standardisation of interfaces between suppliers' sites and buyers' systems. XML has been developed by the XML Working Group (originally known as the SGML Editorial Review Board), formed under the auspices of the World-Wide Web Consortium (W3C). It uses the Internet HTTP (HyperText Transport Protocol) for data exchange, but has a higher level meta-data representation.

The XML syntax is illustrated at the end of Chapter 7. In principle, it enables the standardisation of data structures, including complete blocks of data for a particular purpose, known as documents. Standardisation of the data structures, together with a standard transfer protocol such as HTTP, gives a standard interface that any online trader can follow.

Although XML syntax itself has expressive power only a little greater than that contained in the existing EDI standards, it is set to find application in many areas other than trade data interchange. A document type definition and a style sheet can contain transformation and additional information in transportable form that had to be embedded in a translator for EDI. Thus, it provides a flexibility, portability and ubiquity that will bring with it cheapness of the basic processing engine (to be included in future Web browsers, for instance) and a large skill base. This should ensure that all are able to benefit from it, whereas EDI has only really benefited the larger companies that could afford the translators and skilled personnel, recouping costs through heavy usage. For all its promise, there is still some way to go before XML realises its potential.

9.8 OBJECTS AND MIDDLEWARE

It should be apparent from the case studies in Chapter 6 that integration can be a messy and complex affair. This isn't really news to many software and systems engineers, as they have been coping with the problems of heterogeneity for many years. The holy grail being sought to save the beleaguered integrator is a common backplane, rather like a hardware bus, into which a variety of software applications can be plugged. The reason that this is no mean challenge lies in the sheer diversity of data structures, protocols and architectures that have been used to build systems. They all have to be taken into account, as most real solutions call for data or processing that is encapsulated in some legacy software.

There are two paths to a solution. The first is conceptual—a standard way of defining software objects. The second is more pragmatic—mechanisms for communication between these objects.

Objects

Objects are no more than pieces of closely related processing and data (something tangible like a modem, or something abstract such as an intelligent agent). To count as an object there must be a well-defined boundary or interface that provides a set of methods for interacting with the object (Figure 9.13).

There is more to object-orientation than just interfaces, though. They provide a basis for describing network components in a common framework so that their interaction with each other can be understood. Hence they give a systematic basis for the description and construction of complex (software-rich) systems from simpler components, or objects.

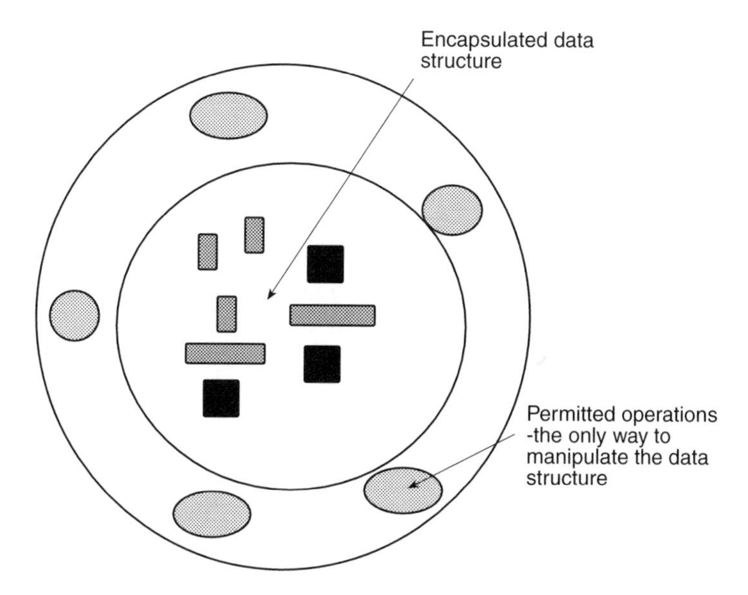

Figure 9.13 An object

The basis of object orientation lies in three simple concepts—encapsulation, inheritance and polymorphism (these slightly pompous terms can be rebadged, respectively, as hiding, breeding and pretending, if it helps):

- *Encapsulation.* This is all about hiding an object's internal implementation from outside inspection or meddling. To support encapsulation you need two things. First, a means to define the object's interface to the outside world. Secondly, you have to have some form of protocol or mechanism for using the interface. An object's interface says 'this is what I guarantee to do, but how I do it is my own business'. It is useful to regard of the interface as a sort of contract—the object promises to provide the specified services and a user promises to abide by the defined interface. The mechanism for using the interface might vary: it could be sending a message or something else entirely—it doesn't really matter as long as the relevant interface conditions are created. The most widely used mechanisms for communicating between objects are outlined below.

  There are two key consequences of encapsulation. The first is one of simplifying the job of users of an object by hiding things that they don't need to know. The second, and more important benefit, is that it is much easier to maintain a degree of consistency and control by prohibiting external users from interfering with processing or data to which they should have no access.

- *Inheritance.* This comes in two closely related but subtly different forms. The first form is 'inheritance of definition'. One of the strengths of object-orientation is the ability to say something like 'object B is of the same type as object A but with the following modifications'. If you are at all familiar with object-oriented terminology, you will probably recognise this as 'object B's class is derived from object A's class'. However it is phrased, the basic idea is that an object's interface can be defined in terms of behaviour inherited from other classes of object, plus some modifications, additions or deletions of its own.

  The advantages of inheritance of definition are twofold. First, there is obvious economy in expressing an interface in terms of a set of already defined portions with a small amount of extra, specific detail. The second advantage is in the lessened potential for introducing errors. It is easier and safer to specify a small set of changes to an already defined and probably working interface than to respecify the whole thing.

  The second form of inheritance flows naturally from the first. It can be described as 'inheritance of implementation'. Suppose you have an implementation of an object of class A and then define a B class as inheriting all of A's behaviour, but just adding one extra facility. It would be very useful if you only had to specify the new function and then the B-class re-used all of the remaining A-class facilities automatically.

This is precisely what happens in object-oriented programming environments.

In a distributed environment it may be useful specify interfaces by using inheritance. However, if an object is at the other end of a network there is no way to tell if its implementation uses inheritance or not. Encapsulation applies here as much as anywhere else!

- *Polymorphism*. The third main object-oriented characteristic is polymorphism. This simply means many forms, and it refers to the ability of an object of one class to be treated as if it were of another class. In object-oriented programming this facility is usually only available to classes related by inheritance. So, objects of class B derived from class A may be able to substitute for objects of class A if they preserve interface of class A as a subset of their own. It is not usually the case, however, that another class, say C, which has precisely the same interface as class A, can be used in its place. This is because the mechanism used to implement polymorphism usually dependent on the shared code base that comes through inheritance of implementation.

  However, things are a little different if we assume a networked situation. If the object at the other end of the wire claims to support particular interface and implements all of the necessary protocols, it is impossible to tell if it is a real object of that type or not. It really is a case of a rose by any other name would smell as sweet. So in this environment, polymorphism is probably more usefully thought of as compatibility.

The idea of objects has pervaded much of the thinking in computing for the past ten years, and it has gradually permeated different part of the software and computing systems industries. Early on, objects were introduced into programming languages and, indeed, languages such as Smalltalk have been highly influential in the development of ideas of object-orientation.

As object-orientation has matured its influence has spread beyond mere programming languages to take in systems design, software development methodologies, requirements capture, maintenance and the whole development lifecycle. Yet more recently, the idea of objects has begun to take hold, not just in the development of software, but also in the run-time operation of computing systems, their evolution and their integration. It has even had an impact on the integration of legacy systems with their newer counterparts: the legacy system may, for example, be encased in an 'object wrapper' to allow it to participate with newer 'pure' object systems.

When we undertake an object-oriented design, the only pre-requisite is really our ability to 'think objects' rather than thinking in terms of more conventional software concepts. However, if we are to implement a system that has recognisable objects at run-time, we need some sort of system components that will support objects as distinct from any other software modules. A conventional computer operating system does not do this: it manages things such as processes, files, I/O devices and so on, but has no notion of objects. Hence, over the years there have been various attempts to produce object-oriented operating systems. Although there seems no particular reason why such an approach should not be feasible, the reality is that no object-oriented operating system has become a serious competitor to other more established, 'conventional' operating systems.

UML

The concept of objects needs to be supported with some language for specification, visualisation, construction and documentation. UML represents a collection of the best engineering practices that have proven successful in the modelling of large and complex object systems—it is the successor to the modelling languages found in the Booch, OOSE/Jacobson, OMT and other methods.

Many companies are incorporating the UML as a standard into their development process and products, which cover disciplines such as business modelling, requirements management, analysis and design, programming and testing. A major attraction of UML is that it facilitates the integration of software development tools from a wide range of manufacturers who conform to the standard. By using a common, UML compliant repository (e.g. the Microsoft Repository), UML models can be exchanged between the different tools. While primarily a standard for object modelling, it is becoming increasingly important for component-based development as it offers the possibility of standardising component descriptions.

UML is becoming the natural choice as a modelling notation for component development based on underlying OO distributed technology. The wide tool support and the exchange of models through the Microsoft Repository are obvious attractions. It is not yet clear how it will scale to describe much larger and complex components, how it will support transaction-based systems or how well support for process flow will be implemented. However, the extensibility of the notation and the weight of the major vendors behind it almost guarantee its leading role.

UML doesn't directly tackle how to address large grained semantic descriptions, or how to express non-functional issues. At the moment these tend to be expressed in plain English. These are easy to understand, but also easy to misinterpret and are open to ambiguity. For the use of complex commercial components to proliferate, such descriptions will need to be tightened up. There are currently no standards or methods defined for doing this. While XML could be a notation for doing this, it isn't necessarily the right one, and it certainly doesn't address the real issues of how to define those semantics and quality standards. The Open

Applications Group (OAG) is working in this area with a view to be able to integrate large-scale applications.

Object middleware

Instead of object-oriented operating systems, the industry trend has been towards the development of pieces of 'object middleware' that can co-exist with a conventional operating system and handle all the object stuff about which the operating system is completely unaware (Figure 9.14). This middleware comes under various names such as CORBA and DCOM but, for the time being, we shall simply refer to it as 'object middleware'. This keeps discussion general and avoids homing in on the solution favoured by a particular manufacturer or standards body (those gory details will come a little later!)

The reasons for the growth in object middleware rather than OO operating systems is probably to do with the fact that, as well as simply supporting objects, the middleware has come to be expected to do some other things:

- The growth of interest in object-orientation has been paralleled by a growth in the demand for distribution. Hence, the object middleware is required not only to handle the interaction of objects on a particular machine, but also to manage the interaction of distributed objects.

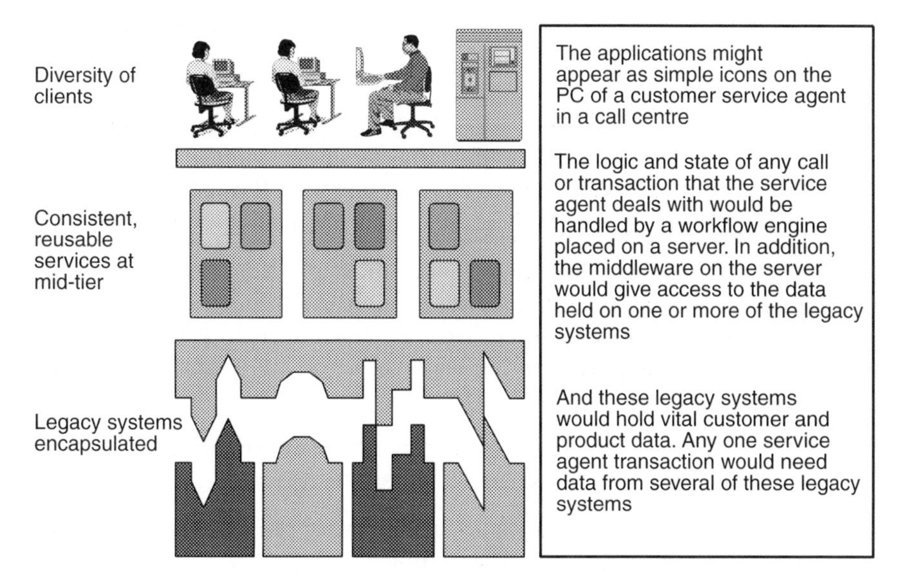

Figure 9.14 Middleware—the glue that attaches old systems to new developments

- The move to client-server and three-tier architectures implies that various kinds of computer must be able to participate in the distributed object world: mainframes, PCs, mid-range servers, and so on. Each of these typically makes different demands on its operating system, and so a 'one size fits all' OO operating system is unlikely to be suitable for everything. It is much more realistic to have appropriate operating systems for the different kinds of computer, but to have common object middleware that can function alongside any of the mainstream operating systems.
- As already noted, an object environment is expected to support the integration of legacy systems. This is because the wholesale replacement of legacy operating systems is not a practical option, and some way of integrating them into new developments is needed (see Figure 9.14). Again, a more practical approach is to implement a piece of object middleware that can support objects alongside conventional software components.

So, if sophisticated eBusiness applications are to be constructed quickly, some form of object middleware is needed. We have given some reasons why there is more to it than simply building an object-oriented operating system, but what, in detail, do we expect our object middleware to do for us?

Functions of object middleware

There are a number of things that we should expect our object middleware to allow us to do. The main things are:

- *Object instantiation*. The ability to add new objects to a system and for it to record information about the objects: their names, their version numbers, and the services they offer, for instance. It must be possible to make objects active (to run their code) and to take them out of service.
- *Object binding*. The object middleware must provide mechanisms that allow objects to be 'bound' together in such a way that they can be used collectively to perform useful functions. This may include the facility for objects to determine (dynamically) what other objects exist in the systems, and what functions they can perform.
- *Object invocation*. The middleware must allow an object to invoke operations on another object and to obtain results back.
- *Transaction services*. The object middleware must provide services similar to those of a transaction processing monitor. It must, for instance, allow operations to be grouped together and handled in such a way that either all the operations within the group are performed successfully, or none

are performed.

- *Data and information management services*. Objects often only exist to process information. The object middleware needs to provide services, for example, for the storage of persistent data and for the handling of such things as documents and multimedia data.

Whilst performing these functions, the middleware is expected to provide the basic transparencies that are inherent in an ideal distributed system:

- Access transparency (the hiding of the details of mechanisms such as communication protocols by which one object interacts with another).
- Location transparency (an object can invoke operations on another object without needing to know the physical location of that object).
- Replication transparency (it should be possible to create a set of clones of an object and for any operation to be performed identically on all members of the set as though it were being invoked on a single object).
- Migration transparency (this takes location transparency a step further, and allows the possibility that an object may move around a system whilst still allowing other objects to make use of their services).
- Failure transparency (in many cases where an object invokes an operation that results in some form of failure, recovery action can be taken and the correct result can still be obtained).

There are other desirable features (Norris and Winton 1996), but anything that satisfies the above criteria goes a long way to easing the construction of 'plug and play' applications.

Object middleware: the main contenders

Enough of the theory. Let's curtail discussion of idealised object middleware. What, in reality, has the industry done towards turning these desirable features into real products?

To understand the different products and their relationships, the first thing we can say is that there are two main contenders for the overall architecture of object middleware. One, CORBA, is the output of a consortium of vendors and object enthusiasts (the Object Management Group), and the other is DCOM, the architecture developed by Microsoft. We shall describe these two architectures, and then try to explain all the other offerings in terms of how they fit with either (or both) of the two.

CORBA

The Common Object Request Broker Architecture (CORBA) is a standard produced by the Object Management Group (OMG) to allow applications to communicate with one another independent of location, platform or vendor. In effect, it is a messaging bus for communications between objects. An early version of CORBA was introduced in 1991, and this defined the Interface Definition Language (IDL) and Application Programming Interfaces (APIs) to enable client/server objects to interact using a specific implementation of an Object Request Broker (ORB). A more recent version (CORBA 2.0) was released in 1995, and this specifies how ORBs from different vendors can interoperate.

CORBA is a client-server middleware that enables a client to invoke a method on a server object, which can be on the same machine or across a network (see Figure 9.15). The ORB intercepts the call and is responsible for finding an object that can implement the request, pass it the parameters, invoke its method, and return the results. The client does not have to be aware of where the object is located, its programming language, its operating system, or any other system aspects that are not part of an object's interface. CORBA also provides a set of service that can be used by developers (naming, event notification, security, persistence, etc.), and facilities. On top of these layers are domain-specific frameworks of business objects. OMG Special Interest Groups and Task Forces have been set-up to agree the standards for these domains.

Primarily aimed at object technology, CORBA is one of several competing sets of distributed technology standards that will provide the plumbing for a true distributed component-based approach. The next release will include support for multiple interfaces (like DCOM), DCOM/CORBA interworking and asynchronous messaging.

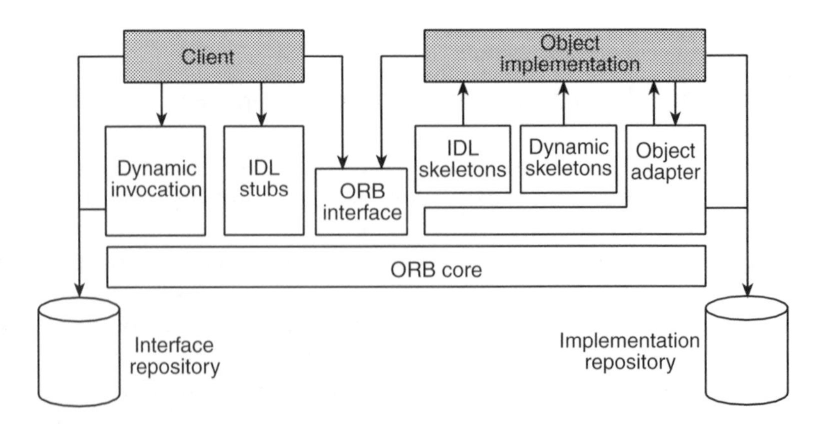

Figure 9.15 CORBA—one variety of software glue

CORBA-II is based on a reference model which consists of the following parts:

- *The Object Request Broker (ORB).* This middleware component is the foundation of the CORBA architecture: it is the thing that allows objects to make and receive requests from each other, whilst providing at least some of the desirable transparencies mentioned above. It is referred to by some authors as an 'object bus' by analogy with hardware bus structures that allow, for example, a number of circuit cards to be plugged into a common backplane 'bus' and communicate with one another. The ORB allows a similar plug-in approach for software objects, as shown in Figure 9.16.
- *Object services.* Rather than integrating all the middleware into the ORB, many of the desirable object middleware functions are specified as services (i.e. objects!) that share access to the ORB along with user-written objects. Such services include those for creating, deleting and copying objects.
- *Common facilities.* These are other services that are shared by applications and are implemented as yet more objects interacting with the ORB. The distinction between Object Services and Common Facilities is that the former are seen as fundamental to the use of virtually all objects, where the latter are useful facilities such as electronic mail that could be used by a range of object types, but are by no means fundamental to the operation of an object system.

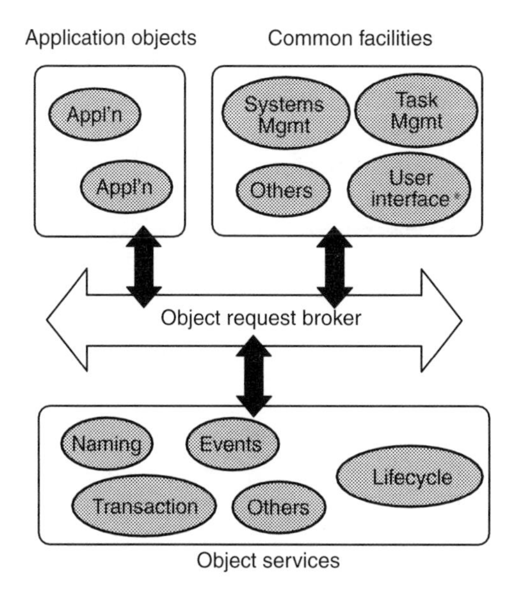

Figure 9.16 The software bus

- *Application objects.* These are the parts that are not specified in detail by the CORBA architecture, but are produced for specific applications. Although CORBA does not specify these components, it has a great deal to say about how they interface to the object middleware (and the ORB, in particular).

  CORBA sees objects as entities that offer one or more services that can be requested by clients. Support for this basic scenario of a client requesting services from an object and getting a result back is a core function of the ORB. However, a simplistic picture such as that in Figure 9.17 conceals a wealth of complexity. First, if we look at the way the client issues its request, there are two alternatives offered by the ORB:

  —*IDL stubs.* In CORBA, the services offered by the remote object are specified using an Interface Definition Language (IDL). IDLs provide a standard way of defining the interface between objects independently of the technology the components will be implemented in. IDLs were first used in the Distributed Computing Environment (DCE) specifications and now both the OMG's OMA and Microsoft's COM have IDLs, although there is no commonality between the three. OMG's CORBA IDL provides a set of bindings to commonly used languages (C, C++, Java, etc.). IDL interface definitions are compiled using an IDL compiler and deployed into a Object Request Broker (ORB) interface repository in the form of stubs which interface client software with the ORB. They appear like a local object to the client software. IDL also creates *skeletons*, which are the same sort of things as stubs, but are located at the server. CORBA also allows a Dynamic Invocation Interface (DII), which allows the connection between client and server objects to be established at run-time. CORBA 2 only allows objects to have a single interface, but inheritance allows sub-interfaces to be specified. COM, on the other hand, allows any number of interfaces.

  —*Dynamic invocation.* An IDL stub as described above is specific to a particular remote object. If the same client needs to request services from a different object, it will need to be provided with another stub

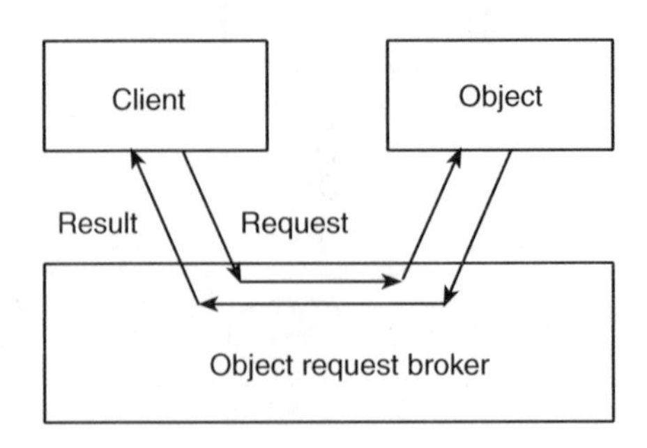

Figure 9.17 Interaction across an ORB

that is tailored for that object. The dynamic invocation interface, on the other hand, is a completely general-purpose interface through which the client can request services from *any* remote object. When calling the Dynamic Invocation Interface, the client must specify (to the ORB) which remote object is to be used, which service is requested, and the set of parameters associated with the request.

Having looked at the two main ways that a client accesses an object via the ORB, we now consider how the object itself handles incoming requests that have been transferred via the ORB. Again, there are essentially two ways that the object can receive incoming requests (and again they represent static and dynamic invocation options):

—*Static IDL skeleton.* We have already mentioned that the services offered by an object are specified in an IDL which can be processed into a piece of code: the client stub. This same IDL can also be processed into a corresponding piece of code that provides an interface between the ORB and remote object. We might think of this as a 'server IDL stub' to complement the client IDL stub. However, in CORBA terminology it is described as a 'static IDL skeleton'.

—*Dynamic skeleton.* As the reader may expect, the dynamic skeleton is analogous to the Dynamic Invocation at the client end. In this case there is no piece of interface code with details of the object's services hard-coded into it. Instead, the skeleton allows the ORB to send the object details of which service has been requested and which parameters accompany the request: it is then up to the object to parse them and deal with them appropriately.

Having seen that there is a static and dynamic option available at both ends of the service request (client interface and object interface), CORBA allows a mix-and-match approach where it is quite legitimate, for example, for a client to make a dynamic invocation to a server that uses a static IDL skeleton.

So far we have treated the ORB as if the only thing it does is to support requests from clients to object implementations. In fact it does a lot more, such as instantiating objects, supplying them with 'object references' (unique identifiers), and a number of other services. These are all private interactions between the object implementation and the ORB. These services are accessed through another interface into the ORB, known as the Object Adapter (Figure 9.18).

- *Distribution.* So far, we have not really introduced the idea of distribution into discussion of the ORB. Indeed, the reader may rather have the impression that the ORB is a single entity that extends across all imaginable machines and allows objects to be plugged in anywhere. In practice, this all pervasive ORB functionality is realised by a number of real ORBs (typically, one per machine) that have to be interconnected by an Inter-ORB

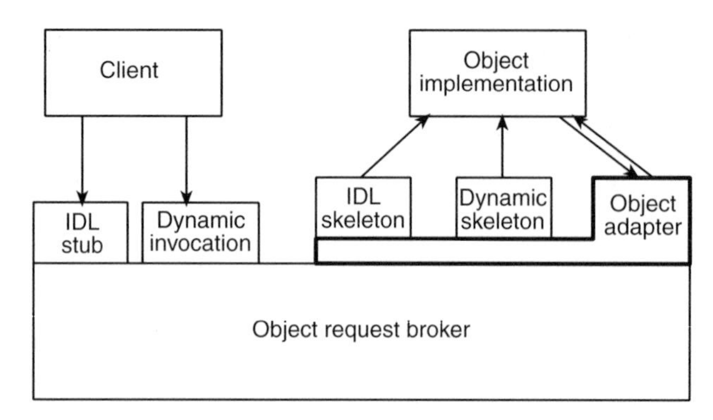

Figure 9.18 The Object Adapter

Protocol (Figure 9.19).

Like most protocols, the Inter-ORB protocol is defined in terms of layers:

Transfer syntax for IDL: CDR
Message Syntax: GIOP
Mapping to underlying transport: IIOP
Underlying transport protocols: TCP/IP

In the above diagram, each of the layers contains the following:

—Transfer syntax for IDL—IDL is the predominant means of specifying data types for CORBA objects. For transfer between ORBs, IDL data

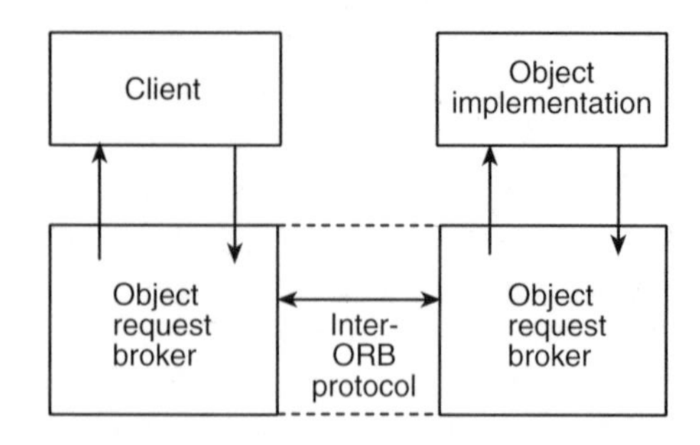

Figure 9.19 Communication between ORBs

types are encoded according to the Common Data Representation (CDR) definition.
—Message syntax layer—General INter-ORB Protocol (GIOP). This defines the seven message formats for all general inter-ORB interactions.
—Mapping to underlying transports—Internet Inter-ORB Protocol (IIOP) specifies how GIOP messages are carried over a TCP/IP network.
—Underlying transport protocols—these are TCP/IP.

Optionally, inter-ORB protocols can be implemented using remote procedure calls. Here, the GIOP/IIOP combination is replaced by yet another protocol called ESIOP (Environment Specific Inter-ORB Protocol).

The source of CORBA, the Object Management Group (OMG), was a consortium of 500 or so companies, including virtually al the main computer and software companies, with the notable exception of Microsoft. They have gone their own way in this area with COM and DCOM.

COM/DCOM

The Component (or Common) Object Model (COM) and Distributed COM (DCOM) specifications cannot be regarded as standards as such because they are the proprietary offering from a single vendor, Microsoft. However, the ubiquity of Microsoft software on the desktop means that client software in a distributed client-server architecture will almost certainly be a Microsoft application, or run on a Microsoft operating system and need to interface with Microsoft applications. Thus, in any large-scale enterprise it is very difficult to ignore the need to interface with Microsoft Object Models.

COM was introduced by Microsoft in 1993 as a natural evolution of the OLE (Object Linking & Embedding) paradigm that they had been using in their suite of Microsoft Office products and the VBX components for Visual Basic. Initially intended for application on a single machine using the Windows operating system, it was expanded to allow communication with components on other Windows NT-based systems, and thus Distributed COM (DCOM) was introduced in 1996. COM 'objects' can in fact be any one of a variety things: a regular C++ object, an OLE component, a piece of code, or a complete application such as one of the Microsoft Office products.

The main characteristic of a COM object is its interfaces. There are some basic interfaces which all objects must implement, but beyond that the user can have any number of interfaces. Somewhat unusually, the objects themselves do not fit the normal objected-oriented model of supporting class inheritance, but it is the interfaces that are inherited. Thus, to create a new interface, an existing interface is sub-classed. Once a COM or DCOM component is complete, it is compiled and supplied as a binary executable for the particular environment for which it is intended.

Microsoft also introduced Active-X controls as components intended for use in Internet applications. Initially these had to be written in C++, but since the release of Visual Basic 5.0 it was possible to write them in VB. As a result they have in fact been primarily used for Windows desktop application. These two facts have led to a considerable expansion of commercially available components. Microsoft is now expanding the model with COM+ which will include transaction processing services, making the components more suitable as a standard for server-side components. COM and DCOM architectures can only be used on Microsoft platforms, so inevitably Microsoft are largely 'going it alone', while most of the rest of the software industry is lining up against them with a converging CORBA and Enterprise Java Beans offering (explained next).

Enterprise Java Beans

One of the key attributes of the introduction of the Java environment is the 'write once-run anywhere' approach. Components in Java are called Java Beans, and were originally intended for delivery over the Internet to run on the client PC. By making use of visual design tools and design patterns, the aim is to be able to rapidly generate new components without writing code.

Most Java Beans were initially produced as GUI components, and didn't lend themselves to server-side operation. The rapid increase in the use of WWW front-ends to business systems created an approach based on a thin-client requiring the development of a multi-tier server architecture and non-visual server components. Enterprise Java Beans was therefore launched to extend the component model to support server components.

Visual Java client components operate with an execution environment provided by an application or Web browser. Java server components, on the other hand, operate in an environment called a 'container'. This is an operating system thread with supporting services that is provided by a major server-side application such as database, web server or Transaction Processing (TP) system.

A class of Enterprise Java Beans (EJB) is only assigned to one container, which in turn only supports that EJB class. The container has control over the class of EJB and intercepts all calls to it. Effectively it provides a 'wrapper' for the EJB, exposing only a limited application-related set of interfaces to the outside world. The appearance of the EJB to the outside world is called an Enterprise Java Beans Object. The container is not an ORB, and client identifies the location of the EJB via the Java Naming and Directory Interface (JNDI). EJB will support CORBA protocols and IIOP. EJB interfaces can be mapped to IDL and can access COM servers.

EJB and CORBA are rapidly converging, and its likely that future versions of CORBA will define the standards for EJB.

Directories and LDAP

Data from eBusiness transactions need to be stored in just the same way that records from paper-based deals are kept. The electronic dual of the filing cabinet is the database, and a merchant's records can be summarised and stored in large databases known as data marts or a data warehouses. To get at this data, which is often stored remotely, some form of common access mechanism is needed, and this is where a directory service with its attendant access protocol plays its part.

A directory is like a database, but tends to contain more descriptive, attribute-based information. The information in a directory is generally read much more often than it is written. As a consequence, directories don't usually implement the complicated transaction or roll-back schemes regular databases use for doing high-volume, complex updates. Directory updates are typically simple all-or-nothing changes (if they are allowed at all). It is common for directories to be tuned to give quick-response to high-volume lookup or search operations. They may have the ability to replicate information widely in order to increase its availability and reliability, without reducing response time.

There are many different ways to provide a directory service. Different methods allow different kinds of information to be stored in the directory, place different requirements on how that information can be referenced, queried and updated and how it is protected from unauthorised access. Some directory services are local, providing service to a restricted context (e.g. a specific service on a single machine).

Other services are global, providing service to a much broader context (e.g. the entire Internet). Global services are usually distributed, meaning that the data they contain is spread across many machines, all of which co-operate to provide the directory service. Typically, a global service defines a uniform namespace which gives the same view of the data no matter where you are in relation to the data itself.

The Lightweight Directory Access Protocol (LDAP), defined in RFC 1777, is a directory service protocol that runs over TCP/IP. It supports a global directory service and is, therefore, very relevant to eBusiness. It has a directory service model that is based on entries, with each entry a collection of attributes that has a name, called a Distinguished Name (DN). The DN is used to refer to the entry unambiguously. Each of the entry's attributes has a type and one or more values. The types are typically mnemonic strings, like 'cn' for common name, or 'mail' for email address. The values depend upon what type of attribute it is. For example, a mail attribute might contain the value mnorris@iee.org. An invoice attribute would contain a textual record.

In LDAP, directory entries are arranged in a hierarchical tree-like structure that reflects political, geographic and/or organisational boundaries. Entries representing countries appear at the top of the tree. Below them are entries representing states or national organisations. Below them might be entries representing people, organisational units, printers, documents, or just about anything else you can think of. Figure 9.20 shows an example LDAP directory tree.

In addition, LDAP allows you to control which attributes are required and allowed in an entry through the use of a special attribute called 'objectclass'. The values of the objectclass attribute determine the schema rules the entry must obey.

An entry is referenced by its distinguished name, which is constructed by taking the name of the entry itself (called the Relative Distinguished Name, RDN) and concatenating the names of its ancestor entries. For example, the entry for Steve West in the example above has an RDN of cn=S.M.West and a DN of 'cn=Steve West, o=BT, c=UK'.

LDAP defines operations for interrogating and updating the directory. Operations are provided for adding and deleting an entry from the directory, changing an existing entry, and changing the name of an entry. Most of the time, though, LDAP is used to search for information in the directory. The LDAP search operation allows some portion of the directory to be searched for entries that match some criteria specified by a search filter. Information can be requested from each entry that matches the criteria.

For example, you might want to search the entire directory subtree below the BT for people with the name Steve West; retrieving the email address of each entry found. LDAP lets you do this easily. Or you might want to search the entries directly below the c=UK entry for organisations with the string 'Telecom' in their name, and that have a fax number. LDAP lets you do this, too.

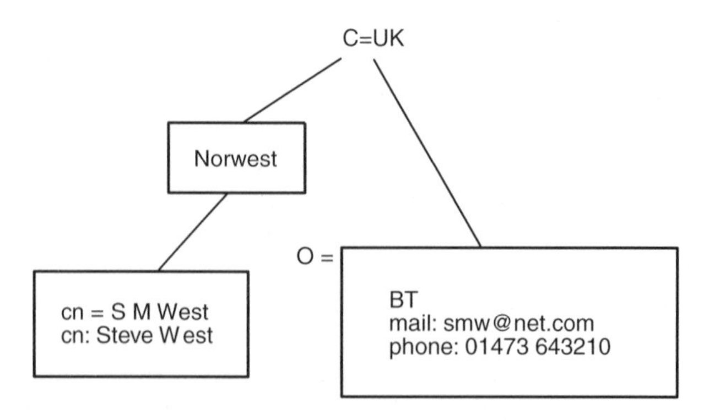

Figure 9.20 An example LDAP directory tree.

Some directory services provide no protection, allowing anyone to see the information. LDAP provides a method for a client to authenticate or prove its identity to a directory server, paving the way for rich access control to protect the information the server contains.

The LDAP directory service is based on a client-server model. One or more LDAP servers contain the data making up the LDAP directory tree. An LDAP client connects to an LDAP server and asks it a question. The server responds with the answer, or with a pointer to where the client can get more information (typically, another LDAP server). No matter which LDAP server a client connects to, it sees the same view of the directory; a name presented to one LDAP server references the same entry it would at another LDAP server. This is an important feature of a global directory service.

9.9 ALLIANCES AND INITIATIVES

With such great things predicted for eBusiness, there are a whole host of organisations seeking to advance, standardise or promote it. They tend to each focus in a specialist area, and in this section we introduce just a few of them. Starting with this selection, it is fairly straightforward to find any of the others.

Telecommunications Information Networking Architecture Consortium (TINA-C)

TINA has a service architecture that is primarily designed to support telecommunications and information management services. The consortium is made up of telecommunications operators and equipment manufacturers, computer manufacturers and software companies. TINA access devices (Computation Objects) interact with other objects via the Distributed Processing Environment (DPE), which provides transparency of location, distribution and replication. Implementations of the DPE are strongly based on the Common Object Request Broker Architecture (CORBA) protocol. TINA includes session and subscription models to support service provision. There have been a number of trial TINA platforms developed, mainly for providing telecoms carrier services, but also including Electronic Shopping Mall, Video on Demand, and Customer Network Management services.

Jini

Jini is Sun's network architecture, which takes the form of a distributed system that behaves as a network federation. Any computing resource (e.g. disk storage, application, etc.) can register itself as a service to any other registered device within the federation. It provides lookup services, a service leasing model, transactions and event models. It is based on Java technology, with Java Remote Method Invocation (RMI) as the inter-object protocol.

Java-based access devices integrate straightforwardly with the architecture, whereas non-Java based applications require a proxy. Several scenarios for the use of Jini have been put forward, most notably around the idea of renting applications.

Wireless Knowledge

Wireless Knowledge is a joint venture between Microsoft and Qualcomm. Its main focus is on ubiquitous wireless device access to central services. Its technology is based on the Microsoft BackOffice family and Microsoft Commercial Internet System (MCIS), and best suits devices that are based on a Microsoft OS (e.g. Microsoft Windows 98, Microsoft Windows CE). Wireless Knowledge's strength is in its working relationships with most wireless carriers in the US, and covers all digital Wide Area Network technologies, including GSM. The first architecture product, called Revolv, became available to commercial end-users in 1999. Microsoft have also described a home networking architecture called Universal Plug and Play, which is based around their Windows 2000 platform. It is intended to be used to interconnect 'smart objects' around the home.

IETF

The IETF have now taken over the OTP protocol from the original OTP consortium, in which Mondex were key players. As Mondex is a true payment only protocol, there need to be higher level trading protocols that provide the environment in which the payment can take place. These higher level protocols cover issues such as negotiation of terms, invoicing, receipting and delivery, etc. As some of the main providers of electronic payment systems (Cybercash, Mondex, Globe ID) were involved in the original OTP specifications, it is likely that the non-payment elements of their systems will migrate to a common set of protocols.

World-Wide Web Consortium (W3C)

In November 1995 the World-Wide Web Consortium produced a working draft (version 0.1) for a MicroPayment Transfer Protocol (MPTP). There has been recent activity in the W3C micropayment group to define a payment API. This work is predicated on the existence of a wallet on the user's client, but with the expected use of XML this could have an impact in that it may define a standard payments tag that can be embedded in the HTML pages that a retailer publishes. This tag would then be interpreted by the client to initiate a payment using an appropriate payment mechanism.

Open Financial Exchange (OFX)

The OFX standard grew out of a joint initiative by CheckFree, Intuit and Microsoft in 1996. OFX bill presentation is a part of the standard which allows corporates to develop a system that can interface with partners and bill consolidators. CheckFree has set up an E-Bill partner program to encourage wider adoption of Internet billing. It currently offers e-billing and payment services to over two million customers in the US and electronic links to 1000 merchants. AT&T joined with CheckFree in mid-1998 to offer residential customers the ability to view and pay their bills over the Internet using CheckFree's electronic systems.

ISRF

The Internet Screenphone Reference Forum (ISRF), whose members are device manufacturers, telecommunications operators and software companies, is developing reference specifications for Internet screenphones. These cover requirements for the devices, applications, core services supported and service management aspects. One specific issue being addressed is event notification, whereby a service can notify the user of the arrival of a message, for example e-mail.

9.10 SUMMARY

In this chapter we've raced through some, but by no means all, of the technology that supports eBusiness. As is often the case, the last thing you know is what to put first, but here we started with a few of the burgeoning range of devices used to transact online business—from Set-Top Boxes to

Kiosks. Given that people need to impress their individuality on all of these devices, we then moved on to explain the main article for achieving this, the smart card.

The next main part of the chapter looked at the various ways in which the user kit can be hooked up. We explained a little of how the Internet works, what the commercial data networks, can do and where mobile access fits in.

Finally, we move on from tangible artefacts to abstract ideas—markup languages, objects and middleware technology, and standards. All of these ideas appear in real life as the software that powers eBusiness applications and services.

The overall aim of the chapter has been to give an awareness of the components that underpin electronic commerce. There are probably many more details lurking inside most practical implementations, so most of the references either fill this detail in or lead to more information that does so.

REFERENCES

A comprehensive set of references to all of the technical areas covered in this chapter would run to several hundred pages. Given below are some of the guides that we have found useful in understanding the breadth of eBusiness technology along with specific references to the concepts and ideas introduced through the chapter.

Access devices

Norman DA, *The Invisible Computer*, MIT Press, 1998.

Bregman M, The convenience of small devices: How pervasive computing will personalise E-Business, *IBM Research Magazine*, **36**(3), 1998. http://www.research.ibm.com/resources/magazine/1989/issue_3/bregman398.html

Internet Screenphone Reference Forum (ISRF), http://ftp.ccett.fr/isrf/

Telecom Law & Policy Resources, *PC-TV Convergence: Set-Top Box Standards*, Blumenfeld & Cohen Technology Law Group, http://technologylaw.com/telecom_policy.html

Braun D, *Coming In 2007: The Wearable Internet*. CMP Media Inc. September 1997. http://www.techweb.com/wure/news/1997/10/1009wearable.html

Smart cards

ISO 7816 Identification Cards—Integrated circuit(s) cards with contacts, http://www.iso.ch.cate/

PC/SC, http://www.smartcardsys.com/
Java Card, http://www.javasoft.com/, http://java.sun.com/products/javacard/
Multos, http://www.multos.com/
Smart Cards for Windows, http://www.microsoft.com/windowsce/smartcard/

Internet technology

Frost A & Norris M, *Exploiting the Internet*, John Wiley & Sons, 1997.

Data networks

Akins J & Norris M, *Total Area Networking*, John Wiley & Sons, 1998.
Norris M & Pretty S, *Designing the Total Area Network*, John Wiley & Sons, 1999.

Mobility

Mougayar W, Web business will soon go wireless, *LANTIMES*, April 1998. http://www.cybem.com/lantime6.htm
Rysavy P, General Packet Radio Service (GPRS), *GSM Data Today*, September 1998. http://www.mobile-systems.bt.co.uk/info/mobiledata/gprs/gprs.htm
Marchent BG, *Development of ATM and IP Networks for Future Mobile Multimedia Systems*, Fujitsu Europe Telecom R&D Centre, July 1998.
Wireless Application Protocol (WAP) Forum home page, http://www.wapforum.com/
GSM, http://www.etsi.fr/smg/

Electronic cash standards

Ford W & Baum M, *Secure Electronic Commerce*, Prentice Hall, 1997.
EMV, http://www.mastercard.com/emv/

Markup languages

West S & Norris M, *Media Engineering*, John Wiley & Sons, 1997.
XML, http://www.w3.org/XML/, http://www.oasis-open.org/cover/

Objects and middleware

The Common Object Request Broker: Architecture and Specification, Revision 2.0, July 1995, updated 1996.

Orfali R, Harkey D & Edwards J, *The Essential Distributed Objects Survival Guide*, John Wiley & Sons, 1996.
Norris M & Winton N, *Energise the Network: Distributed Computing Explained*, Addison-Wesley, 1996.
CryptoAPI, http://www.microsfot.com/security/tech/cryptoapi/
Cryptoki, http://www.rsa.com/rsalabs/pubs/PKCS/html/pkcs-11.html
CORBA, http://www.omg.org/corba/whatiscorba.html
DCE, http://www.osf.org/dce/index.html
UML, http://www.omg.org/news/pr97/umlprimer.html, http://www.rational.com/uml/index.jtmpl

Alliances and initiatives

Wireless Knowledge home page, *wireless***knowledge**™, http://www.wirelessknowledge.com/
TINA-C home page, http://www.tinac.com/
Jini Product home page, Sun Microsystems plc, http://java.sun.com/products/jini/
UN/EDIFACT Standards Database (EDICORE), http://www.itu.int/itudoc/un/edicore.html
RosettaNet, http://www.rosettanet.org/
OBI, http://www.openbuy.org/obi/about/OBIbackgrounder.htm
W3c, http://www.w3.org/TR/REC-xml

10

Who is Going to Make Money out of All of this and How?

The best way to predict the future is to invent it

Alan Kay

It is a pleasant sunny afternoon. You are driving into the country for a few days vacation, when the announcement system in your car advises you to pull over. It says that you need to get to a garage because the head gasket is set to blow sometime in the next fifty miles or so. A map appears on your Routemaster telling you exactly where to go to get it fixed. You press the button to confirm that you are on the way. By the time you arrive at the garage, a qualified mechanic is ready and all the parts have been delivered from the supplier. An hour later you are back on the road.

Far fetched? Well, it may not be the way of the world just yet, but there is no real reason why this scenario isn't an everyday event. After all, engine management systems can detect impending failures and relay messages to a control point, the Global Positioning System (GPS) makes it possible to find the nearest garage, and the electronic business technology explained here could get the right person with the right tools and parts for the job. So, the technology is no barrier (http://www.symbian.com).

Furthermore, everyone wins. Your car doesn't blow up and leave you stranded on your first holiday weekend for ages, the garage gets some passing business, the mechanic gets some welcome overtime and the car manufacturer can ensure that approved parts are used for preventive maintenance. Even the breakdown services win—they can concentrate on more challenging problems when the routine failures are catered for.

If we already know how to make minor magic happen, surely there is major magic just over the horizon. We are most probably in the last few years of the domination of the industrial economy. Very soon the information economy, or network economy, will take its place as the main engine of wealth.

In this chapter we will look at some of the underlying trends that are driving this revolution, map out a bit of the landscape, and offer a few guidelines for success in the new world. Often these guidelines seem counter-intuitive and against conventional commercial wisdom.

First, though, let's look at some of the emerging market dynamics that are beginning to appear as online business starts to flourish.

10.1 NEW DOG, NEW TRICKS

There is no doubt that customer and business relationships are being re-formed by Internet technologies. Distance is less of an issue, the 'information worker' is now a part of the scene and competition is global. All of this threatens previously profitable industries, whilst creating a multitude of new ones.

Early on in this book, we introduced West's law of eBusiness. This stated that for every enterprise that attempts to make money on the Internet there is an equal and opposite enterprise that offers the same thing for free. In many instances, this law holds true. There are, however, times when we need to delve a little deeper.

There is an emerging counterpoint to West's law. This is Norris's law, which states that nothing is really free on the Internet—it's just that it is paid for in what we traditionally regard as an unconventional way. By way of illustration, in conventional commerce, the model is one where I provide you with goods or services and you pay me money or some other valuable item. In Internet-land things are much more complex, and there are an increasing number of examples that illustrate Norris's Law. For instance,

- Freeserve is a 'free' Internet Service Provider (currently growing at about 100,000 users per month) which, despite making no direct charges to its customers, receives its revenues from Energis, the long-haul network company for the phone calls that those customers generate.
- Geocities provide 'free' hosting of Internet web pages. That is, as long as you carry compulsory advertisements on your pages. This, in effect, funds the service.
- BrainBox, who host a 'free' online market. You make money by selling part of the online market that you have established (i.e. the brand) to a

third party who is, to all intents and purposes, a franchisee.

- BT Argent, a micropayments site which charges a small commission on each transaction from traders whose products range from beer guides to the Monica Lewinsky story.

In each case, the service provider offers something that is free to the consumer, but they still manage to recoup some money. The mechanism by which they earn their crust is indirect (and different in each instance). There are probably many more ways of getting some return out of the Internet. The point, though, is that the relevant models are not traditional ones. The dynamics of the market and the trade it supports are different, and will doubtless evolve. In the next sections, we look at some of the main drivers and issues for the next era.

10.2 FROM COMPETITION TO CO-OPERATION

The traditional and well-established models for commercial activities draw heavily on the strategies developed for war and sport. Competitive advantage is gained by being stronger, faster or leaner. When this is applied in an environment with the speed and flexibility of the Internet, there is very little stability, and organisations are prone to sudden extinction as their market changes. A new way of looking at business will be needed—one based on co-operation rather than competition.

Supply networks

Just as networking stand-alone PCs transformed business processes within organisations, the networking of stand-alone companies will transform business processes between organisations. Greater cooperation between organisations will be built using Extranets and Community of Interest Networks. The ability to exchange information with ease will revolutionise the way in which organisations interact with each other, just as it has done within organisations that have deployed an Intranet (Norris, Muschamp & Sim 1999).

Initially, this networking will lead to more efficient automation of supply chains. However, as the technology becomes more pervasive, it will permit organisations to work much more collaboratively with suppliers and customers, and also with organisations that lie outside the traditional supply chain. This will lead to the formation of supply networks where not only is an organisation sharing information and working collaboratively with its immediate suppliers and customers, but also with a much wider range of organisations (Project ION). The Internet portal sites, such as Excite and Yahoo, are an example of this mode of commerce (Excite, Yahoo). Many of the functions of these sites (i.e. stock quotes, travel, classifieds, dynamic maps, etc.) are provided by a network of partner firms.

Supply chains were originally formed to increase business efficiency. After all, it is not practical for an individual to order a single chocolate bar directly from the manufacturer. The cost to the manufacturer to process each order, deal with returns, ship the item and arrange billing for individual customers led to roles for intermediaries, such as distributors and retailers.

This is changing with the ease and low cost of communications provided through the Internet. For some goods the classic supply chain is already being bypassed. Goods that pose no logistic problems (such as music and software) are obvious examples. More recently, others have followed the same path, including Virtual Vineyards, a wine mail order group. By organising many growers on their web site and utilising efficient logistics, Virtual Vineyards have created a simple supply network (Virtuel Vineyards).

As experience and knowledge of the new ways of doing business grows, the relationships and the electronic 'bonding' between the organisations involved in supply networks will grow stronger, and it will become increasingly difficult to draw the boundaries between these organisations. We will soon reach a stage where rather than having companies competing against each other we will have supply networks competing against each other.

The virtual corporation

The concept of the virtual corporation has been envisaged for some time, but now it is a practical reality. Such organisations are dynamic and can be configured at will. Resources, whether people, information or machines, can be called on when needed. They lay outside the small management core and comprise the 'exo-firm' (Murphy 1996). Many consultancy companies are organised in this way. To fulfil a client's task the appropriate experts are assembled into a custom-company. If the client demands a lot of specialist expertise or a presence in many countries, it can be recruited. The appropriate people are found, they execute the task and then disband when it is finished.

There is none of the inertia found in highly structured, hierarchical companies. In a hierarchy it is assumed information is scarce, and therefore members are ranked so that orders are followed. Virtual corporations are networks. They rely on communications and the kind of *ad hoc* organisation that the net allows. They are well suited to dealing with change.

Customers become employees

As information technology becomes part of our everyday life, we notice that we can now do many tasks that previously required specialists. The obvious example is desktop publishing. We now create our own stationery, newsletters and even greetings cards. Often we will do the design part (personalisation) and let the specialist produce the goods. Technology gives us the tools to design our own products.

This trend continues online. Dell's 'Buy a Dell' site lets customers design a machine to their own precise specification. Once the new computer is designed (i.e. amount of memory, hard disk size and other key features are chosen), the customer hits the 'buy' button and the component supply, assembly and logistics functions are invoked to supply the product in a just-in-time manner. No employees are needed to talk to the customer or to enter the customer or product details, and this reduces Dell's costs. As we take responsibility for more and more roles in the buying process, the distinction between employee and customer becomes blurred.

10.3 POWER TO THE PEOPLE

The eBusiness revolution changes the customer's relationship with a company. Customers become more powerful as a result of sharing information. This trend is then amplified by the progressive new organisations who, in an attempt to win customers and loyalty, pander to these powerful customers' needs.

Buying communities

A new form of intermediary is emerging to help to organise buyers in order to increase their buying power. They are often called 'virtual communities', and they act as a broker between buyers and products (Hagel & Armstrong 1997). These communities act as the 'champion of the customer', helping the customer to get the best deal, and they provide merchants with access to informed customers who are eager to buy. They usually operate in particular niches, and in many ways are the natural, online manifestations of the traditional magazine. Advice and discussion forums are offered to ensure customers fully understand the product domain, and are certain about a buying decision.

To make a purchase we must first identify a need, then we analyse products and merchants, we negotiate the best deal, and finally, we review our purchase. Buying communities understand and augment this buying behaviour.

When product domains are complex and marketing messages from manufacturers are confusing, then a buying community site is very valuable. They can offer the buyer help to understand and choose the most suitable product (PersonalLogic). They can compare merchants' prices (Acses) and can offer a potential customer reviews and feedback from previous customers (Amazon), who are just like themselves (FireFly). They can even tell the customer the precise profit margin the merchant is making (Auto-by-Tel). With such powerful advice, customers will be sure of a purchasing decision and can make informed purchases with confidence. A good example of a site that offers all of these tools is DCResource.com, a buyers' advice site for digital cameras (DC Resource).

The key to making this service work is to instill a culture of feedback and review amongst customers. For a short period, when Amazon began, they offered a sizable reward for the best book review that was submitted every day. This information 'primed' their recommendation system. It now contains enough review content to persuade customers that they will in fact like a book, and other related books, because similar people to them like these books.

This trend has threatening implications for big brands. In general, we resent exaggerated and confusing marketing messages, but we often buy these products because we can't find better, impartial advice. Big brands reduce buyers' doubt, but not as well as buying communities can. We will expect to see products compared honestly, and the complaining letters as well as the glowing testimonials. Big brands cannot work like this, it doesn't seem to make commercial sense to fairly compare a competitor's product on one's web site. Big brands will lose customers whose loyalty will be won by impartial buying communities that they can trust.

The customer is king (again)

It is in a company's interests to encourage their customers to talk to each other. By offering discussion facilities, customers will exhibit a smarter, well informed, collective intelligence. What these customers consume is then often produced by themselves—the 'audience are the authors'. This group of customers can help the company push its technology further, finding ever more creative ways to use the company's products (Dutta 1999). They can help each other to solve problems, and can provide early problem reports. Software companies have long used enthusiastic users like this to do beta testing (see the many Microsoft news groups dedicated to individual Microsoft products).

These users are often the most knowledgeable, often more knowledgeable than a company's own staff, and the resources these customers create can form the equivalent of online user guides and helpdesks. Not only do these customers, help existing customers but they will create new ones by enthusiastically advocating a product or service. By providing the means for customers to organise in this way, a company can turn a previously disparate set of individuals into an easy-to-reach, loyal, niche market.

Power to the purchaser

eBusiness is shifting the balance of power during a transaction towards the customer. By using the net to organise, customers are effectively creating 'Chambers of Commerce for Customers'. By sharing information, experience and details of previous transactions, customers can increase their buying power by acting as a single, larger group. In the longer term, it is likely that people's agents will work together to make large buying communities. A small piece of software could link together millions of people into a single market.

Traditionally, a customer can have little influence on a firm. If a firm is powerful then a complaining letter or call is unlikely to cause any great effect—each customer is a lone voice. Even email, though much easier to send than a letter, can simply end up in the electronic waste basket or get a worthless automatically-generated response.

Public discussion groups change the balance. A single customer's views can now be seen by all. This vastly increases a company's incentive to act. It is now common for customers to search for news articles about a firm and its products. Customers can see what others think before deciding to buy—it's like having hindsight in advance! Dejanews, the news group indexer and search engine, makes this very easy (Deja News). There are many news groups dedicated to particular companies and products, specifically to allow customers to pass on their good and bad experiences.

This is such a powerful means to regulate companies' behaviour, and to inform consumers, that it has been suggested that public discussions forums are made mandatory, and are managed by government watchdog organisations (Andersen 1999).

10.4 ADVERTISING

The nature of advertising will change. A type of new-media marketing will supersede broadcast advertising. In this, individual promotional efforts can be precisely measured in terms of success and increased revenues.

Traditional advertising is a very imprecise activity—half the money spent is wasted, but no-one knows which half!

There are survey techniques to estimate a campaign's effectiveness, but there is no way for a company to measure the exact return on investment of any single promotional effort, or what exactly stimulated a buyer to make a purchase.

By contrast, the net affords many ways to measure the effectiveness of advertising. When a buyer hits the 'buy' button, the advertisement, special-offer, or link that brought the buyer to the purchasing decision, can be recorded. The power of any advert to induce a sale will be measurable almost immediately. Advertisers can thus spend their promotional resources only in the places where results are good, so advertising costs will no longer be fixed, but rather performance-based.

How might a company manoeuver into a position to exploit precise targeting and performance-based pricing? This is a role for yet another role—the web advertising and promotions intermediary.

Customers are becoming increasingly aware that they are being tracked, logged and watched by companies who are eager to learn about them. Learning about customers makes it easier and cheaper for firms to sell. Therefore, customers' profile information has a value. High quality, accurate data (given willingly by the customer) will become an extremely valuable commodity. It therefore seems likely that customers will demand some payback for the valuable information they surrender about themselves. An intermediary can manage customer's profiles and ensure that customers receive a reward for it.

As companies only pay for advertising based on success, they are able to make savings and can offer buyers better prices. The prices for the buyers will need to be genuinely and provably cheaper, and this saving is, in effect, the buyers' payment for surrendering their profiles. CyberGold has pioneered the early instances of this.

New agencies and companies will form to serve this market, and will threaten traditional big-media advertising firms (using TV, radio, print). The net, and then interactive TV, will become powerful commercial mediums, defined largely by advertisers' needs (rather than by the breaks in the major TV soap operas). We will quickly move beyond banner adverts and the other old-fashioned techniques hastily adapted for the web.

10.5 TRUST ME, I'M A TRADER

Companies will need to work hard to build and keep their customers' trust. They will have to tell customers what information is being stored, what it will be used for, and who it will be shared with. If customers are confident and happy with this they will begin to trust the company.

Standards for describing privacy practices and ratings schemes for companies' privacy procedures are appearing (P3P, Website trust). Trust will be built over time, over the course of many transactions, and will have a mutual value to both company and customer.

Our privacy may be traded for goods or discount. In exchange for a year's free access to an online music site, customers may allow the music site to log their listening habits. They may receive junk mail as a result, but they will have saved the subscription fees.

Trust can also be an emergent property of a community of trading partners. After a transaction, both parties rate each other which then modifies their respective trust level. Over time a reliable rating level can be calculated. The auction marketplace run by eBay uses such a system to allow members to judge the trustworthiness of potential trading partners based on their past transactions.

10.6 PARADISE LOST

Clearly the eBusiness revolution will lead to business casualties. In this section we outline the characteristics that expose traditional businesses to danger from new players.

Threatened industries

Which industries are threatened by the revolution in eBusiness? As we have observed throughout this chapter, eBusiness thrives on an incumbent's inefficiencies. Industries that are founded on an old style of intermediation and that extract large margins from the transactions that they broker are fertile regions in which new, highly efficient players will grow. Only businesses where the traditional or quaint nature of its product is a unique selling point can survive without change.

Some industry sectors will undergo an entire restructuring. The travel industry, commodity retailing and personal finance sectors are already well on their way. New players will attract customers with innovative and cheap services, and will use this body of customers to bypass the current dominant players (PriceLine). These changes will be rapid and brutal. The growth rates for some (and attrition rates for others) will be spectacular. The classic eBusiness case study, Amazon.com, exemplifies this threat. In little over three years it had become a multi-billion dollar book, CD, video and drug retailing behemoth.

Other industry sectors may not face such a thorough restructuring, but the marketing, customer management and back-end processes of virtually every firm will be affected. If companies try to resist the inevitable tide, they will be submerged. In fact, if they don't seize the initiative and act quickly, even if this will threaten their existing profitable business, they may still go under—it's adapt or die; and survival is optional.

Increasing returns

In many businesses, there is a concept of diminishing returns—beyond a certain point, the cost of increasing capacity exceeds the profit gained from the increase. By way of contrast, the digital economy is linear and the more you sell, the more you earn (Wired). Increasing returns cause one player to grow, very rapidly, to dominate a market space. Microsoft was an early example of an increasing returns business. The more its products sold, then the more likely it was that new customers would buy its products so that they could inter-operate with existing users. Amazon.com is another increasing returns business. Its massive database of customer opinion and review means that customers are more likely to buy via Amazon because they will have the best chance of making a good, well-informed purchase. It may be difficult to become established, but once you are, business can be a virtuous circle!

Increasing returns causes the biggest to get bigger and to lock out all other competitors (which will increasingly arouse the attention of fair trade bodies). Understanding this growth dynamic is very important—it is vital to move fast and to grab market share as there is often no second place.

Of course, no party goes on forever. Changes in the market will come about that allow another fast growing upstart to gain a foothold. Also, a dominant player will need to maintain their position—the real problem with being a success is that you have to go on being a success.

10.7 AT LAST—1984

For many years we have had concerns about 'Big Brother' and what information is known or can be deduced about us. Using the net exacerbates this. Everything we do can be stored and recalled later—our behaviour can be traced. A few decades ago we could be anonymous but now, with the advent of credit cards, mobile phones and the web, organisations can tell what we have bought, where we have been, who we have spoken to and what we have seen.

Crimes are often solved with the aid of credit card records. Imagine how easy it will be to trace individuals when almost every interaction will leave an electronic record. It may be possible to analyse a person's entire lifetime. This would lead to many interesting possibilities in the future if the necessary historical data were available: history could be taught in schools by using data-mining techniques to create case studies of real people; crimes could be solved many years after the event as data-mining techniques become more powerful.

Is the threat from Big Brother real? It seems unlikely that governments will have the will or wherewithal to do this. Instead the threat comes from businesses—the massive corporations that will have access to enough data. Our defence against such abuse will come from the enhanced customer power afforded by the net. Companies that transgress the new economy's protocols of trust and fair trade will be exposed, and will be severely damaged.

10.8 SUMMARY

We have shown that new online and communication technologies have sparked a revolution in the way that we do business. This revolution is no illusion or passing fad—it will endure and will have far-reaching effects, many of which are yet to emerge. This chapter has considered some of the ways in which the business world will change and adapt in the information age.

In the vanguard of this new age are the entrepreneurial start-up CEOs who, armed with potent new business models, and preaching attractive new consumer philosophies, are attracting customers in droves. The technology that allows them to fulfil their imaginings and realise their dreams is all there, but as in any revolution, there will be casualties, and in this instance it will be the sluggish or inefficient firms who are up against the wall. The main text can, we hope, guide the aspiring new age traveller to a place of enlightenment and safety.

REFERENCES

Acses, A shopbot for books. http://www.acses.com

Amazon.com, http://www.amazon.com

Andresen T, Consumer power via the Internet. *First Monday (http://131.193.153.231)*, **4**(1), 1999.

Auto-by-Tel, http://www.auto-by-tel.com

CyberGold, http://www.cybergold.com

DejaNews, http://www.dejanews.com

Dell, http://www.dell.com/store/

DC Resource, http://www.dcresource.com

Dutta S, Success in Internet Time. A Global Information Exchange report. www.infoexchange.bt.com 1999.
FireFly, http://www.firefly.com
Excite, http://www.excite.com
Project ION, http://www.labs.bt.com/people/callagjg/ion/supply.htm
Hagel J & Armstrong AG, *Net Gain: Expanding markets through virtual communities.* HBS Press, 1997.
Murphy S, Corporate metamorphosis: the effect of new media. *First Monday* (*http://131.193.153.231*), **1**(1), 1996.
Norris M, Muschamp P & Sim S, The BT Intranet: Information by Design. *IEEE Computer*, March 1999.
PersonaLogic, http://www.personalogic.com
PriceLine, Online travel demand aggregation, http://www.priceline.com
Website trust, rating organisation, http://www.truste.com/
Yahoo, http://www.yahoo.com
VirtualVineyards, http://www.virtualvineyard.com
Wired, http://www.wired.com/wired/archive/3.10/arthur.html
P3P, Platform for Privacy Preferences, http://www.w3.org/P3P/

Appendix 1: Case Studies

Business neglected is business lost

Daniel Defoe

For all the models, technology and standards that we have built up and explained in this book, eBusiness is essentially all about making money. This ultimate goal can be thought of in its constituent elements—market share, efficiency and tunover—but there is no getting away from the fact that the proof of a pudding in its eating. And this means checking that eBusiness actually delivers competitive advantage.

This appendix describes some actual examples of eBusiness being translated into profit. Many of the technical details presented in the body of the book can be laid over each of the following. The point, though, is not how these organisations implemented eBusiness, but rather that they did so to good effect. In each case, they had to understand the advantage that they could achieve by moving into an electronic market.

A1.1 CASE STUDY 1—FEDERAL EXPRESS

Scope

Fedex is the world's largest express transportation company. It has over 137,000 employees worldwide, and handles 2.9 million items every single day across 212 countries.

FedEx, like every other major business, has its competitors. The United Parcel Service (UPS) is Fedex's biggest rival, with each of them aiming to take the express delivery service away from the 'incumbent', the US Postal

Service. Fedex has 43% of the market against UPS with 27%, with Airborne at 18% and DHL at 3% as the smaller rivals.

For years, FedEx and UPS have been going 'tit for tat' in a battle to offer customers more delivery options, at lower cost, with greater convenience and reliability. For the future, however, both are gearing up to become full service logistics providers that specialise in orchestrating the flow of both goods and information between customers and suppliers—and this means ebusiness!

Expanding core business

Both FedEx and UPS have put in place sophisticated tracking and control schemes to properly manage the massive logistics needed to run their worldwide operations. One of the services developed to get one step ahead was first put in place by UPS, who allowed customers to have electronic access to the tracking systems, in order for them to see where their packages are at any one time.

However, FedEx asked the question 'How many customers really want to track a package which is being delivered overnight anyway?' They built on the UPS idea by creating an end-to-end service, rather than enhancing the capabilities of the middle-man delivering the goods.

So to the FedEx Virtual Order (FVO), in which the corporation plays a part in all aspects of the transaction between customer and supplier. This new service is aimed at suppliers of goods requiring FedEx's delivery service, and was launched with the slogan 'Increase the size of your business without increasing the size of your business'. FVO (Figure A1.1) works as follows. The supplier creates an online catalogue of its products on FedEx's 'Merchant Server'. Using the World-Wide Web, customers have access to the product catalogue and can browse it online. Any orders are electronically submitted and the FedEx shipping label is automatically produced. The package can then be put together and collected by FedEx for delivery. The supplier's inventory system can also be linked to the merchant server, if required. The original UPS idea is still present, as tracking of the package can be done online by both supplier and customer.

Advantages for the supplier include access to 18 million customers worldwide, automation of order taking and other 'lower-level tasks' and a 'free' presence in the electronic marketplace.

The pitch for this service is mainly at the suppliers rather than their customers. The vast majority of FedEx's customers will still want to order goods in the more traditional way, but this gives suppliers a real opportunity to get their products 'onto the Internet'. In addition, the supplier is getting a fully automated order management service which they can use to replace the more traditional methods, as business grows using this service.

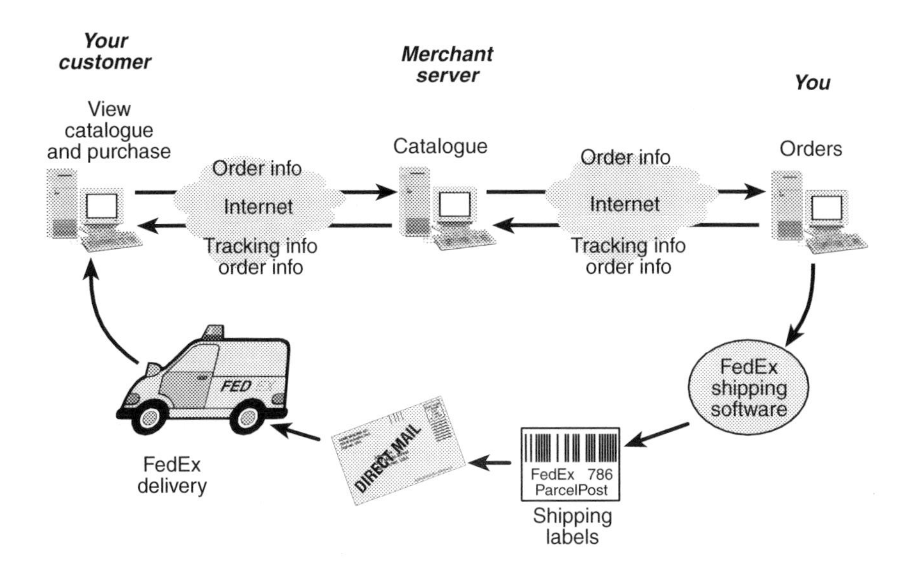

Figure A1.1—The Federal Express Virtual Order set up

From the customer's perspective they are getting a seamless service from order request through to fulfilment, via a single interface using a single medium.

This is a good example of a fully integrated eBusiness solution, implemented using basic Internet-type technology.

Why does it work?

There are a number of reasons why FVO has been such a success, and has increased the FedEx market share. Other companies have tried to use the Web as the basis for displaying and selling their wares, but have simply treated it as another 'brochure medium'. The vast majority of Web ordering ends in customers calling a phone number to place the final order. Fred Smith, FedEx CEO describes this as 'commerce interruptus'.

A lot of companies want to cut the costs tied up in their inventory, but the effort often requires a radical redesign of their internal processes. FedEx trains its sales people to help suppliers examine their internal supply processes so that they can reduce their inventory to sales ratio by using FedEx to implement just-in-time delivery.

The real beauty of the system, however, is that by providing the facility for an end-to-end service, both the supplier and customer are tied to FedEx for the order fulfilment.

A more deep seated reason for the success of FVO is that FedEx realised twenty years ago that they were in the information business. They have stressed that knowledge about a cargo's origin, whereabouts, destination, estimated time of arrival, price and cost of shipment was as important as its safe delivery. To take advantage of this, they have insisted that a network of state-of-the-art information systems (including laser scanners, bar codes, software and electronic connections) be put in place to work along-side their transport networks. Their competitors can of course put the same technology in place, but it is the first in the market that wins the game.

FedEx have taken a look outside the traditional bounds of what a transport company would understand as its business. It has embraced more of the value-added aspects of order request and fulfilment, such as helping suppliers market their products, hosting web servers, and by providing a secure and trusted route for the flow of information about the order and its delivery.

However, FedEx have also realised that their core business (that is, moving packages and parcels around) will always be required. In the not-too-distant future, intelligent software purchasing systems might be able to find the exact match between customer requirement and the final product, cutting out the middle-man, but there will still need to be an efficient means of getting the product from A to B.

In short

FedEx took advantage of the value of the information underlying their core business. They built on it by getting into their customers' internal processes and offering them advantage whilst locking them into their core business

A1.2 CASE STUDY 2—CISCO SYSTEMS

Scope

Cisco Systems is one of the world's leading suppliers of routers and other network equipment. Like FedEx, they operate in a highly competitive market, with companies such as Bay Networks and Newbridge keen to grab some of Cisco's market share. The high value nature of Cisco's business mean that it is vital to meet customer requirements first time and every time.

Quality of service

Being part of the communications community, Cisco have been in the vanguard of Internet use. Over the last few years they have augmented the basic software download capabilities long offered over the web with more advanced services such as trouble-ticket and service order tracking, bug tracking, question and answer forums, and an array of configuration utilities and intelligent troubleshooting engines.

These services have helped Cisco ring up billions of dollars in electronic sales over the Internet. They have also helped Cisco customers to save equally impressive amounts by increasing the speed and accuracy of orders placed.

With regard to configuration utilities, for example, Cisco uses a product configuration package developed by Calico Technology to help its systems engineers, value-added resellers, and large customers accurately configure complex, build-to-order products such as routers and hubs. Authorised users access the configuration page on the Web, search for products, choose a particular model, and specify the options they need.

A Cisco 7000 router, for example, can be configured for such things as software options, power supply, power cables, memory, route and switch processors, and interface modules. Error messages indicate if a particular equipment configuration is valid, and suggestions are provided to help users arrive at the best configuration. The bottom of the page has controls that allow users to send the order to Cisco. Many large customers report significant savings due to the accuracy of the system. Indeed, one large telecommunications company saw its ill-specified orders fall from nearly 70% to virtually zero through use of the system.

In addition to placing orders, Cisco customers can also view the status of their service orders, right down to the Federal Express tracking number for shipments, with hot links to the shipping company's package-tracking page (see the previous case study!).

Cisco offers a Web-based support service called the Bug Toolkit, which allows customers to search a database by keyword or bug ID and create one or more 'Watcher Bins' in which updates on selected bugs can be stored. Agents can be set up to search for bug updates that meet certain criteria and store those alerts in a Watcher Bin. Updates are presented to the user on the home page, and also can be emailed or faxed. The update frequency for email and fax can be set to 'notify on new alert' or 'summarize alerts weekly'.

Cisco also provides an Open Forum area where customers can ask questions that can be answered by other customers or Cisco personnel. There is a special forum available where people who have achieved the Cisco Certified Internetwork Engineer (CCIE) rating can chat and assist one another in real time. Cisco also offers a troubleshooting engine, which asks the user questions in an attempt to zero in on the solution to a problem.

Impact

Cisco has made a major commitment to the Web as a way of doing business. This commitment actually extends well beyond service and support to include Web-based order and invoice tracking, product configuration and pricing. The company now attributes 70% of its annual sales revenue to the Web, which translates to over US$5 billion per annum (or about US$15 million a day).

In addition to capturing sales, the site saves the company around US$300 million per year in operating expenses, which it can plough back into research and development. Figures from Cisco customers are more difficult to confirm, but the general perception is that the level of checking, speed of fulfilment and consistency of problem tracking that can be achieved with the eBusiness approach saves a considerable amount of user time, and hence cash.

A1.3 SUMMARY

The final concept we introduce in this book is the fear and greed model. An organisation may be drawn to eBusiness as a way of making its products easier to buy. The old marketing adage that the best way to sell is to make it easy to buy applies. A more subtle drive to introduce eBusiness would be fear of being left behind by the competition. It is clear from the first two case studies that eBusiness didn't just help the companies to make more money, it also distinguished them as leaders. In each case, others have deployed similar services.

FURTHER READING

Irving L, *The risks and rewards of electronic commerce. Journal of International Banking and Commerce*, 1998. http://www.ARRAYdev.com/commerce/JIBC/9801-3.htm

Kalakota R & Whinston AB, *Frontiers of Electronic Commerce*, Addison-Wesley, 1997.

Lauder G & Westall A, *Small Firms On-line*, Institute for Public Policy Research, London, 1997.

Appendix 2: The Gods of Technology

The real question is not whether machines think but whether men do

B F Skinner

This appendix picks up on a couple of points that recur through our book. The first is that eBusiness is as much (if not more) about business as it is about technology, and the second is that transferring ideas from one discipline can solve problems in another.

In the authors' experience, one of the most critical factors in any successful project is to match the technology and processes being deployed to the people that have to use them. This means that you have to understand both what is appropriate and when. Of course, some gifted individual can get this right, every time. Mere mortals, though, usually need a little help in making sensible choices.

This appendix builds a model that matches different styles of technology to different styles of organisation. In doing this, we build on the seminal works in the world of management theory—Charles Handy's *Gods of Management* (Handy 1991). An outrageously short summary of this classic book would be that it characterises different types of organisation according to their function and behaviour. From this humble basis, he illustrates what to expect, how to plan and what is appropriate for each type. Handy's analogy brought the specialist area of management theory into the wider public domain, illuminated many aspects of it, and made a complex area accessible to a much wider audience.

Having lived with both computers, communications and various people for many years, the authors believe that eBusiness developments are amenable to the same sort of analysis that Handy has applied. After all, it is organisational need that these computers and networks are being used to satisfy, so it is not unreasonable to expect the technology to have been created in the image of its master. To explore the analogy we don't have to dive into all of the fine technicalities that abound—our aim will to be place it into a framework that helps every man to cope with otherwise overwhelming detail.

A2.1 EXPLORING OLYMPUS

Most organisations have spent the last thirty years or so coming to terms with computers and the networks that connect them. Over time, the technology has come to be more and more important to the business: the computer that was once a novelty now holds vital business information and processes.

Quite often, these information networks[1] have grown with the business that they serve. along the way, they have taken on many of the characteristics of their creator: a centralised organisation with a central mainframe, an alliance of home workers with their Internet connections, and so on. So, if we look at the distinct types of organisation, we are likely to find parallel technical characteristics. Without further ado, then, here are the four technically-minded Gods revered by the computer network mortals.

- *Zeus*, of all the Greek gods, was the one who wielded the most personal power and influence. He was metaphorically (and in many cases, literally) the 'father of the gods', and his power extended out via the lesser gods to affect the lives of men. Charles Handy characterises a Zeus figure as one who wields power from the centre of a web, and operates through a 'club culture'. Here we are looking at an organisation with a strong control centre. Everyone knows where the company headquarters are, and that all information is directed at headquarters (because information is power).

  The technology that supports this type of roganisation is mandated. Typically, it would be provided by one trusted supplier. This supplier would, most likely, be looking after all of the computer and network requirements for the organisation. This works because the supplier has very clear requirements and is happy to serve such a powerful God. The organisation, meanwhile, wants to get on with the real business and doesn't want to be too bothered with incidentals like technology. The credo that pervades the organisational web can be paraphrased as 'no-one ever got sacked for buying IBM'.

[1] By which we mean Intranets, Virtual Private Networks, Enterprise Networks, and any other network of computers that deliver inforation to members of an orgainisation.

- *Apollo* was the god of music and poetry though, in ancient Greece, these were arts that complied with very strick rules of structure and form—it would have been unthinkable to write a line of poetry that did not scan! Likewise, temples of Apollo epitomise classical architecture, and Handy characterises the management in such organisations as being constructed and as being concerned with roles and responsibilities. Everyone knows the part they play, and there are rules and regulations to cover just about every eventuality.

  To keep this sort of organisation running, there has to be strict control over what can and cannot form part of the technical infrastructure. The diversity of function within the organisation means that no one supplier can satisfy all needs, so some sort of systems constraints are needed. More often than not, this type of organisation will have an internal group whose role will be to assure order and compliance to something called an 'IT strategy' or 'Network Architecture'. This group will say what can and cannot be used. They will police what is purchased, and will analyse the fit between the organisation's strategy and the evolving marketplace. The credo here is that 'we must control diversity'.

- *Athena* was a hunter and was primarily concerned with getting a result! The organisation that Handy associates with Athena runs on a 'task culture' in which teams bring a range of specialist skills to focus on particular tasks, and the results and 'deliverables' that are produced. This culture is likely to see information technology in terms of a set of projects rather than as a steady-state management job. It is likely to be dominated by computer scientists and developers.

  As the key to this sort of organisation is the skill of the individual, there tends to be significant focus on the best tools for the task. Small groups within the organisation would have a fair degree of autonomy in what they do—after all, they are the experts and the company relies on them for its market edge. Not surprisingly, there are many suppliers into this type of organisation. There is no constraining strategy, though, and diversity is rife. The developer has such a strong part here that the central control of Zeus or the delegated control of Apollo will not work. Over time, the pile of superseded technology grows, and the costs of this weeks wonder tools spiral. Fortunately, the expertise of the organisation is in demand, and the costs of constant change are sustainable.

- *Dionysus* was the Greek god of winemaking, and is associated with personal freedom. In Handy's analysis, Dionysus takes the Athena model one step further. Instead of the task being the focus of attention, it is the individual. What is really important is the capabilities of technology to meet the individual requirements of the end-user.

  Again, this is an organisation that relies on the expertise and imagination of the individual. In fact, the reliance is so complete that few controls are imposed, and what gets done tends to be driven by what can be done.

As well as having many suppliers, this type of organisation will have a considerable amount of home grown technology. The rules are, to a large extent, made by the individual. The credo in this case is (to paraphrase Bismark) 'the art of the possible'. As a rule, this approach work well in business areas that are evolving quickly, where technical capability is a differentiator, and where it is not established just what can be done.

If each of the above stayed separate, in its own niche, this might all be interesting but of little consequence. In the real world, though, things change: organisations grow, merge and change their focus. As this happens, their people, practice and technical infrastructure all need to change.

So we need to understand how each style of computing fits with another, how we move from one style to another, and how to choose an appropriate style. The trouble is that there is an element of competition among our Gods of Technology. Each is resistant to the others. They have each come to a particular view of technology that guides behaviour, and when conflicting ideas crop up, religious anti-bodies gather to repel the new idea (even when it is right).

A2.2 THE GODS IN ACTION

It is one thing to describe something, another to recognise it in the context of everyday life. To provide some illustrations of the gods of technology we now sketch out a brief pen-picture of each, taken from our own experience in the IT industry. We have certainly seen all of these over the years, sometimes within one organisation, sometimes not. At least one of them should ring a bell.

Zeus

A few years ago we visited the IT department of a small mobile telecommunications company. Although the IT manager was certainly a Zeus, the company itself was organised very much on Apollonian lines. Their business was essentially one of buying air time from one of the large cellular networks and selling mobile phones together with a service package. At that time, mobile phones were still in their early days and the prices were high. Hence, the level of service and the quality of the service organisation had to be excellent: if a customer called in with a query about their account it was essential that the service assistant who handled their enquiry had instant access to all the necessary details. Hence, the IT manager of the company had been charged with installing a comprehensive information system that could deliver any information about customers' accounts to terminals operated by the service assistants.

The way he went about this was to approach a number of suppliers who could potentially offer suitable systems, and proceed to assess them through discussions with them and their existing customers. When we talked to him about the selection process that he employed, we (coming from a strongly Athenian culture) culd not really understand that most of the 'evaluation' took place in restaurants or on golf courses not, as we would have expected, in the laboratory or the computer room.

The result was that rather than trying to evaluate the optimum technology, he had selected a company whom he trusted and with whom he could work in partnership. That is not to say it was always a 'cosy' relationship with the supplier. In a particular instance where their core product lacked a feature that he required, he succeeded in mobilising opinion from a range of other customers of the same supplied and bringing pressure from several directions at once to force the supplier to develop the features he required.

Technically, the solution was quite unusual for this sort of application. It was essentially a very large document image processing system. Every paper document (customer contract, customer letter, invoice, receipt, etc.) was scanned into the system. When a customer telephoned in with an enquiry, the service assistant could bring up on his or her screen an image of all or any of these documents.

When we saw the system it was fully installed and working extremely successfully. The Zeus approach had resulted in a very effective system for the company: furthermore, it had been deployed very rapidly. Our own experience of such projects had been one where a great deal of time and teamwork are usually devoted to the evaluation of competing technologies to find the 'best' technical option—a process that is usually fairly drawn-out in its timescale, and expensive. A second advantage to the Zeus approach was that the company required only a very small IT department: most of the technical skill was bought in from the supplier. Furthermore, the close relationship with the supplier resulted in their putting a lot of work into ensuring that the system tryly met requirements. This differed from some of our other experience of supplier relationships (mainly in Apollo IT departments) which have centred on costs rather than trust.

On the down side, the IT manager was perpetually in conflict with the other management in the company, who were classic Apollonian and shared little common ground with him. Even when he had delivered the finished system, they gave him little credit for the success (he was 'lucky' to hit on a supplier who has done his job for thim). Also, whereas his role was key in the supplier selection process and during the deployment, where he was constantly intervening personally to fix some problem or other, the system is now working smoothly, and here is no longer a

Zeus-type role to be fulfilled. The system could more comfortably be brought into the Apollonian structure.

Apollo

Somewhat more in keeping with the author's training and background is a recent project on which they collaborated. This was the development of an online reference that had to be delivered via an organisation's private Intranet. The idea was that it would give several thousand engineers fast and well structured access to a wide range of technical reference material. It was intended to contain guidance on the company's preferred technical options, and to indicate where a particular solution was mandated.

The customer for the work was very clearly an acolyte of Apollo. Expected outcomes were very well defined, with acceptance criteria stated at the outset and a number of 'quality gates' to be achieved during development defined. In addition to the structure imposed on development, there was considerable focus on how the finished product would be used and maintained. Issues such as operability, usability and performance were part of the brief, and both had to be demonstrated as acceptable before the work was signed off.

For this particular project, where the challenge was to organise and efficiently deliver large files to a widely dispersed audience, working for Apollo had a definite plus side. With acceptable levels of technical and process quality laid down, it was clear what needed doing and in what order.

Less clear was how to deal with the unexpected. A notable aspect of working in this type of environment was that deivations from specification or procedure were often not entertained—even when they appeared to be useful. Almost anything outside of agreed scope had to be escalated and getting change agreed often took far more time than it would have been worth. Hence, some of the opportunities that only arise in the course of work were passed over. Of course, this may have been the right thing as it sometimes only takes a few misplaced on-the-fly decisions to undermine a well thought out plan.

One aspect of theproject that only became clear as we neared delivery was that the real users (the developers who would follow the online guidance) were not one and the same as the 'customer acceptance team', who would conduct the initial product trial. In this organisation, the IT strategy was set, enacted and policed by a team of specialists. The end users (who, it turned out, constituted an Athena organisation) tended to follow the rules when they liked them and argue for change when they didn't. Hence the handover of the product into maintenance was drawn out with many user-initiated change requests. Once they had tweaked the gift from Apollo to their liking, it was accepted as something worth having, quite open to new technology as long as it has first been tested and shown fit for purpose.

Athena

In the above example, Apollo might have been in charge, but Athena had significant influence. The authors have worked on a number of projects where the balance has swung right over to the side of Athena, with everything led by technical specialist.

The evolution of one particular network provides a case in point. In this instance, the idea was to exploit existing line plant (copper wires, switches, etc.) by delivering a range of new services over it using sophisticated transmission techniques.

In the absence of any precedents, the approach on the project was to deploy as many of the technical options as possible, evaluate them and choose a preferred set. Before this could be done, some agreement on how things should work together was needed, and there were no established standards. So a variety of interfaces, protocols and data structures had to be specified.

The potential capabilities of the network led a number of related project areas to become involved. In particular, several of the network application teams offered their terminal equipment as a test of the network's capability. duly, a trial was set up and some impressive services were demonstrated.

By this time, the project was accorded considerable importance and was supported by sales, marketing and commercial representatives. The speculative input from these areas (not surprising, given the novelty of the project) meant that the main focus stayed on things technical. For some considerable time, the project delivered a range of improved technical solutions with the release of a commercial service held in abeyance until a clear winner was established.

It was only when standards were established for the network and terminal equipment became available that the technical experimentation stopped. Very quickly thereafter, the nature of the project changed, and many of the concerns and procedures of Apollo came to the fore.

In its heyday, this project was an island of pure Athena. The technical specialists dominated and the quest for a better mousetrap went on without too much deference to where they should be placed or, indeed, whether there were any mice to be caught. They technology all worked in isolation, but as a commercial proposition it wasn't sustainable!

In this instance, the organisation was reliant on technical capability, and so it looked to its specialists for a lead. The specialists did their job until the continual innovation ceased to be viable, whereupon Zeus came to the fore and decided to deploy a standardised product from a supplier. the Athena part of the organisation moved on to build services on the network that they had done so much to bring into being.

Dionysus

Our final illustration is taken from a collaborative project that we undertook with a firm of management consultants. They were a classic Dionysus organisation—a network of experts with little discernible hierarchical structure—and they had an IT department to match. Each consultant required IT facilites that best met the needs of the particular consultancy task on which he was working: individuals chose their own equipment and applications: the job of the IT department was to deliver and manage them. Personal computing was about half-and-half Apple Macinotshes and Windows PCs; a great variety of desktop applications was in use: graphics, presentation software, spreadsheets, small databases, project planning tools, flow-charting, often with a mix-and-match approach using different vendors' products.

The requirements for data integration were quite small, the main requirement being simply to exchange files between machines and packages—not normally a problem nowadays, even when using a diverity of systems. As the company grew, requirements for centralised information systems arose from the need for sharing of information between indivieuals. An early requirement was for their IT department to provide them with a centralised file server through which they could share files between personal computers. Their solution was a Novel file server which adequately supported their need for a diversity of machines and network connections. They they required some more structured tools that would support individuals working in groups. Their choice was Lotus Notes which again placed no real constraints on their individual diversity, whilst providing an added level of groupware across the company. In other words, their 'central' IT systems had to cope with the diversity that individuals wanted: the alternative of adopting 'standard' products was too high a price to pay.

In this company, the IT department was very much at the beck and call of the user. Their IT manager was essentially someone who managed the technicians—not someone who set any overall IT strategy for the company.

A2.3 FINDING YOUR GOD

In reality, there are few instances of a mono-God culture: most large organisations have a mix of two or more. Nonetheless, there is usually a predominant style that could or should apply, and it is important to establish what it is (as opposed to what it is assumed to be).

In truth, it doesn't take very long to establish the prevailing deity—if you ask the right questions. To illustrate, many large organisations believe that they have an information and systems strategy that ripples down from the top. The strategic direction is set by the Chief Information Officer (CIO) and the systems architecture is set by senior IT managers who vet top-level designs before they are implemented by project teams. However, the reality is often quite different. It is the authors' experience that most of the key design decisions are actually taken by comparatively lowly software engineers—not by their managers, far less by the CIO. How do we justify this assertion? Well, first we need to elaborate on what we mean by a key decision. Essentially, it is one that has a large long-term financial impact: a decision that critically affects the success or failure of a project; a decision to include, or exclude, a particular point of flexibility in a design.

As a topical example at the time of writing, most people would agree that the specific choice of data type to represent a real worl dentity such as a date, would be a decision made by a programmer, not a board-level business strategist. However, with the millennium and other date-related software bugs at the forefront of many people's mind, the issue of whether a 'year' is represented by an unsigned integer, a short integer, two ASCII digits or four ASCII digits, is of critical importance to businesses. It probably represents the most costly single element of the information systems budget for many companies around the year 2000.

The Millennium question—or the more general issues of who spotted that there would be a problem with the date format used in computer systems—is often very revealing. In many cases, as explained above, it sometimes reveals a schism between Gods, and highlights the fact that the intended and actual views of an organisation's technical operations are not always the same.

So the question, along with its consequential allocation of blame and responsibility, is a worthy test. As a result and bearing in mind the subtleties explained above, the answer gives us a fix on the God to follow. If a supplier identifies the problem and accepts that it is theirs to fix, then Zeus is Lord. If the lead in finding a solution lies with a specific part of the organisation, all are charged with fixing the problem and blame is not imporotant, then it is Apollo in charge. Athena is at the forefront when the problem may (or may not) have been spotted locally, and it is the fault of the software developers. As far as Dionysus is concerned, the Millennium is a problem for the end user of technology, so it is probably of little technical interest and will receive scant attention.

There are a couple of more general test questions that reveal which of the Gods is really in charge. The first is, 'What sort of computing environment do you have, then?' Archetypal answers would be as follows:

Zeus	'We run IBM'	The supplier is seen as the prime driver. The organisation sets requirements.
Apollo	'Our applications are all DCE compliant'	Conformance to a defined architecture or standard. People are trained to use it.
Athena	'All the online and multimedia technologies'	The latest technology. The development community drive choice.
Dionysus	'We have a range of finance applications'	Best of breed tools from a range of suppliers. Users integrate them.

The second, follow-on, question (and one that relates very much to the local situation) is, 'Who decides what you buy?' In this instance, the case in point might be where a hardware upgrade or new piece of software came from. So, when the question of where you get a couple more yards of disk storage[2] from, the answers would usually be something like:

Zeus 'It's not an issue—we get all of our kit through the company supply scheme. You just pick up the phone and tell them what you need. I've no idea whether it is the cheapest or the best you can get but you get it when you need it. Once you've bought something, it is added to the maintenance contract'.

Apollo 'There is a company portfolio that you can select from. Once you have decided what you want, you have to write a business case to justify purchase. I'm pretty sure that you get a good deal, but it usually takes a whole. Our operations people look after installation and maintenance'.

Athena 'More often than not, we don't have to buy commercial solutions. We can download some utilities off the Internet and use locally developed routines to do just what we want. I'm not sure whether anyone looks after stuff after I've finished with it, though'.

Dionysus 'It's an individual decision. You just have to sort out what you need. For my part, I have a budget and I buy what I need to do the job. I cannot let technology get in the way of the task so cannot afford to shop around'.

[2] Disk storage is consumed in great volumes by modern computing applications, and is usually bolted on as a commodity when more data has to be accommodated.

The above analysis provides an introduction to the four Gods, and gives some idea of which one is prevalent in your own organisation, or that of an organisation you rely on.

So, now we have established who the predominant God is. That is a good start but, as indicated earlier, it is not the whole story. The question, then, is whether this is appropriate to the job in hand. After all, each God of Technology has both strengths and weaknesses. Traditional followers of Zeus might find that there is a wrong place or a wrong time to worship. In the next section, we consider some of the warning signs.

A2.4 FOLLOWING A FALSE GOD

If technology is well suited to the job it is used for, it can be a great enabler. Computers and networks have, in many areas of our everyday lives, had a huge impact: cars use less fuel, thanks to on-board engine management systems; cash is available from Automatic Teller Machines whenever you need it, and many new telecommunication services are now possible.

Just as there are many instances of successful deployment of technology, there are examples where it didn't work. It would take another book to detail the cases in which technology proved to be a non-productive sink for large amounts of cash.

For now, suffice it to say that the common cause in these examples was not the technology *per se*, but the way in which it was handled. In other words, the organisation was not appropriate—they were following a false God.

As a rough guide, Table A2.1 summarises the strengths, weaknesses and most appropriate applications for each of our Gods.

There is a nightmare scenario for each of the Gods of Technology. Since Zeus leaves technical strategy (all, or in part) to a trusted supplier, it is possible to find that the chosen ally fails (or misbehaves). The history of IBM and their erstwhile supplier, Microsoft, comes to mind. With technology so much a part of business success, the Zeusian approach is rather like giving your soul to the devil . . . for safe keeping!

For Apollo, the worry is that a glorious battleship is constructed only to be sunk by holes beneath the waterline. In practice, these would be a strategy that does not move fast enough to accommodate a new technology that blows away the *status quo*. Even if sound, the Apollonian solution may tend to maximise the use of assets, rather than support the people doing the business, so could end up like a well run hospital with no patients.

The failure mode for Athena is simply that so much time is spent at the leading, bleeding edge of technology that the business haemorrhages money. The organisation ends up as the best equipped bankrupt in town.

Dionysus is most likely to fall into complete chaos. The very innovative

Table A2.1

	Zeus	Apollo	Athena	Dionysus
+ve	Fast, responsive	Reliable, predictable	Optimised	Innovative
−ve	Single point of failure	Slow to change—all analysis, no decision	Expensive to run, over-engineered	Dependent on experts
OK	When speed counts and deals need to be struck	In steady state where cost matters	In a growing area where expertise counts	In emerging area where rules are few
e.g.	Retailers, Small businesses	Telecoms, Banking	Stock dealers, Software houses	Online games

sparkle that helped the organisation to make its mark has caused a loss of focus, and the business is awash with ideas, none of which quite work together. All value is vested in a few heads who go on the explore somewhere else.

The examples given here could be a bit pejorative, as there are just as many examples of type within one sector of industry as there are across sectors. Nevertheless, the popular image in each case does typify the characteristics of the God in question. The fundamental point being made is that each God has characteristics that suit one business, but not another.

Achieving a match is important. This may simply be a matter of business as usual. More often in these fast-moving times, it should be a conscious choice that often calls for a departure from the norm. This is easy to say and difficult to achieve for the simple reason that many technical ideas are sufficiently subtle to lull you into believing that you understand them when you have only seen part of the picture.

A2.5 SUMMARY

In this appendix we highlight a problem. It is that the diversity of computing and network technology that underpins eBusiness is stretching the specialist at the same time that its power and application is being deployed by those with many other things on their mind.

We conclude that a simple framework is needed into which technical options can be fitted, but that there is no 'one size fits all' approach: rather, the framework needs to fit the business and the culture of IT within an organisation.

In keeping with previous, successful instances where solutions can be found in history, the idea of the 'Gods of Technology' is introduced as a basis for our framework. It features:

Zeus—centralised, supplier-led
Apollo—diversity control, strategy‡led
Athena—task oriented, developer-led
Dionysus—individual, user-led.

The strengths and frailties of each are illustrated, on the premise that all organisations are bound by their ability to deploy technology, and success here depends upon appropriate homage to the right Gods.

REFERENCE

Handy C, *Gods of Management*, London, Business Books, 1991.

FURTHER READING

Jay A, *Management and Machiavelli*, London, Hutchinson Business, 1987.
Handy C, *The Empty Raincoat*, Hutchins, 1994.
Porter M, *Competition in Global Industries*, Harvard Business School Press, 1986.
Norris M & Winton N, *Energize the Network*, Addison-Wesley Longman, London, 1996.
Ohmae K, *The Borderless World*, Collins, London, 1990.
Drucker P, *Post Capitalist Society*, Butterworth-Heinemann, Oxford, 1991.
Atkins J & Norris M, *Total Area Networking*, John Wiley & Sons, 1998.

Appendix 3

For the snark was a boojum, you see

Lewis Carroll

GLOSSARY

It is probably apparent through this text that eBusiness is a fast moving area driven by both technical and business innovation. Hence the language of the software engineer, the banker, the network designer, the market maker and the information technologist all have their place. Here we draw on these diverse disciplines by listing many of the key abbreviations and concepts in eBusiness. Some of the terms are explained in the main text, many are not. In either case, the aim is to clarify the more complex ideas by giving some idea of their context and application.

Accounts Payable — A business area within a company where selling organisations send invoices, and from which the responding payment is processed.

Accounts Receivable — A business area within a company that generates an invoice for products or services sold by the organisation and processes the responding payments. Also known as billing and collections.

Addenda Record — A record of the Automated Clearing House payment formats that contains payment-related information. This information, sent in standard formats, accompanies payments exchanged through the Clearing House Network.

Active-X	Microsoft technology for embedding information objects and application components within one another. For example, an Active-X button can be embedded in an HTML page that is displayed in a browser window.
Address	A common term used both in computers and data communication designating the destination or origination of data or terminal equipment in the transmission of data.
Algorithm	A group of defined rules or processes for solving a problem. This might be a mathematical procedure enabling a problem to be solved in a definitive number of steps. A precise set of instructions for carrying out some computation (e.g. the algorithm for calculating an employee's take-home pay).
ANSI	American National Standards Institute. The national standards body for the United States. ANSI, through its accredited standards committees, keeps the standards for all applications of technology and mechanics for US industry.
API	Application Program Interface—software designed to make a computer's facilities accessible to an application program. All operating systems and network operating systems have APIs. In a networking environment it is essential that various machines' APIs are compatible, otherwise programs would be exclusive to the machines in which they reside.
Application	The user task performed by a computer (such as making a hotel reservation, processing a company's accounts or analysing market research data).
Application program	A series of computer instructions or a program which when executed performs a task directly associated with an application such as spreadsheets, word processing, database management.
Applications software	The software used to carry out the applications task.
Architecture	When applied to computer and communication systems, it denotes the logical structure or organisation of the system and defines its functions, interfaces, data and procedures. In practice, architecture is not one thing but a set of views used to control or understand complex systems.
Artificial Intelligence (AI)	Applications that would appear to show intelligence if they were carried out by a human being.
ASC X12	Accredited Standards Committee, X12. It comprises government and industry members who create Elec-

	tronic Data Interchange (EDI) standards for submission to ANSI for subsequent approval and dissemination.
Asynchronous data transmission	A data transmission in which receiver and transmission transmitter clocks are not synchronised, each character (word/data block) is preceded by a start by and terminated by one or more stop bits, which are used at the receiver for synchronisation.
Asynchronous Transfer Mode (ATM)	ATM is a standard for high speed fixed size packet communications. It provides a basis for multi-service networks—those capable of carrying voice, video, text etc.
Automated Clearinghouse (ACH)	A network of offices through which banking transactions affecting more than one financial institution are routed, to debit and credit the correct financial institutions.
Automation	Systems that can operate with little or no human intervention. It is easiest to automate simple mechanical processes, hardest to automate those tasks needing common sense, creative ability, judgement or initiative in unprecedented situations.
Authentication	Authentication is the process of verifying the identification of the true sender of a message, and also that the text of the message itself has not been altered.
Bandwidth	The difference between the highest and lowest sinusoidal frequency signals that can be transmitted by a communications channel, it determines maximum information carrying capacity of the channel
Back office	The back office of a financial institution is made up of employees responsible for (1) recording and maintaining the official records of the financial institutions and (2) processing transactions entered into by the financial institutions or its customers.
Batch processing	Batch processing is the transmission or processing of a group of payment orders and/or securities transfer instructions.
Blanket Order	A contractual agreement between trading partners covering all purchases of a specific item or all items of a specific nature for a period of performance (generally one year). Once a Blanket Order is issued releases are made against it, with no Purchase Orders.
Browser	A browser is a computer program that facilitates locating and displaying information on the World Wide Web (e.g. Netscape Navigator or Microsoft Explorer). The browser could work on the Internet or through internal information management systems called Intranets.

Bug	An error in program or fault in equipment. A hangover from the days when an insect caught in an early electro-mechanical computer caused it to fail.
Catalogue	These are central to eBusiness. A catalogue is the electronic equivalent of a shop's shelves, goods, departments, etc. It is the online representation of what is 'for sale' (or more correctly, what is available for trading). The way in which they are constructed is explained as are the processes for keeping them 'live' This can range from a set of web pages and a simple script that allows orders to be taken through mid-range catalogue products that are characterised by a pre-defined structure of product categories and sub-categories, up to large scale corporate catalogues that are customisable and usually feature back-end integration with inventory, stock control and ordering systems. Catalogues for buyers and sellers are different—the former is a virtual catalogue through which the buyer can see competing products from a number of suppliers, and the latter is a structured set of information that represents what a supplier has to sell.
Certificate Authority	A trusted third party that authenticates that a public key belongs to a specific, registered party.
Circuit	An electrical path between two points generally made up with a number of discrete components.
Circuit switching	The method of communications where a continuous path is first established by switching (making connections) and then using this path for the duration of the transmissions. Circuit switching is used in telephone networks and some newer digital data networks.
Clearance	Clearance is the process of transmitting, reconciling, and in some cases, confirming payment orders or security transfer instructions prior to settlement, possibly including the netting of instructions and the establishment of final positions for settlement. In the context of securities markets, this process is often referred to as clearance.
Client	An object which is participating in an interaction with another object, and is taking the role of requesting (and receiving) the required service.
Client-Server	The division of an application into two parts, where one acts as the 'client' (by requesting a service) and the other acts as the 'server' (by providing the service). The rationale behind client/server computing is to exploit the local desk top processing power leaving the server to govern the centrally held information. This should

	not be confused with PCs holding their own files on a LAN, as here the client or PC is carrying out its own application tasks.
Commodity Codes	Codes used to define specific types of commodities across selling organisations. No existing standards exist, and many third party organisations market their proprietary commodity coding schemas.
Commodity Goods and Services	Tangible products and services whose specifications are pre-set by the manufacturer for sale to a broad market. These transactions generally account for over 80% of purchasing activity, and can be sourced from preferred selling organisations.
Configuration	A collection of items that bear a particular relation to each other (e.g. the data configuration of a system in which classes of data and their relationships are defined).
Connection oriented	When information is exchanged over a fixed link with predictable characteristics. The public switched telephone network exemplifies this type of network.
Connectionless	When information is dynamically routed across a network in self contained units. X25 is a widely known connectionless service. This type of network is characterised by possible loss, delay or reordering of information. It is the user who has to implement the end-to-end protocols to reorder, resequence and recover.
Consolidated Invoice	When multiple Orders are issued to a single selling organisation, that selling organisation may return a Consolidated Invoice which aggregates the information of the multiple orders over a pre-specified time period. See Invoice.
CORBA	Common Object Request Broker Architecture. An evolving framework being developed by the Object Management Group to provide common approach to systems interworking.
Customer Inquiry	A requisitioner or buying organisation's request for information, expediting, order change, technical literature, or other help related to an order.
Cyberspace	A term used to describe the world of computers and the society that gathers around them. First coined by William Gibson in his novel *Neuromancer*.
Data	Usually the same as information. Sometimes information is regarded as processed data.
Data compression	A method of reducing the amount of data to be transmitted by applying an algorithm to the basic data source. A decompression algorithm expands the data back to its original state at the other end of the link.

Database	A collection of interrelated data stored together with controlled redundancy to support one or more applications. On a network, data files are organised so that users can access a pool of relevant information.
Database server	The machine that controls access to the database using client/server architecture. The server part of the program is responsible for updating records, ensuring that multiple access is available to authorised users, protecting the data and communicating with other servers holding relevant data.
DBMS	DataBase Management Systems—groups of software used to set up and maintain a database that will allow users to call up the records they require. In some cases, DBMS also offer report and application generating facilities.
DCE	Distributed Computing Environment. A set of definitions and components for distributed computing developed by the Open Software Foundation, an industry led consortia.
Detached digital signature	PKCS #7 signatures can be detached or self-contained. A detached signature is a signature separate from the content to which the signature applies whereas in a self-contained signature the signature and the content are linked. A detached signature is a PKCS #7 data object of type SignedData with the SignerInfos field containing signatures on external data and the ContentInfo field empty. OBI/1.0 specifies the use of a detached signature.
Digital certificate	A document signed with a digital signature that states that a specified public key belongs to someone or something with a specified name. Within OBI, certificates are signed or issued by a trusted third party (Certificate Authority) to requisitioners, servers, and authorized signers of order documents.
Digital signature	Digital signatures utilise public key cryptography and one-way hash functions to produce a signature on the data that can be authenticated and which is difficult to forge or repudiate. A digital signature is often referred to as an encrypted message digest. Within OBI, it is possible to include a digital signature with an order or order request using the 'PKCS #7: Cryptographic Message Syntax Standard' defined by RSA Data Security.
Direct	Purchases made for products and services that are going into the buying organisation's end-product. Generally linked to inventory and replenishment systems,

	and also generally not subject to sales taxes in the US.
Distributed computing	In a move away from having large centralised computers such as minicomputer and mainframes, and bring processing power to the desktop. Often confused with distributed processing.
Distributed database	A database that allows users to gain access to records, as though they were held locally, through a database server on each of the machines holding part of the database. Every database server needs to be able to communicate with all the others as well as being accessible to multiple users.
Distributed processing	The distribution of information processing functions amongst several different locations in a distributed system.
DNA	Distributed iNternet Applications architecture. The Microsoft view of how internet based software should operate. DNA consists of business process, storage, user identification and navigation elements. It builds on established COM , DCOM and Active-X ideas and provides middleware for distributed applications.
DNS	Distributed name system. The method used to convert Internet names to their corresponding Internet numbers.
Domain	Part of a naming hierarchy. A domain name consists of a sequence of names or other words separated by dots. Also, a part of a network
eBusiness	A term that embraces all aspects of buying and selling products and services over a network. The essential characteristics of eBusiness are that the dealings between consumers and suppliers are online transactions and that the key commodity being traded is information. eBusiness is the gateway to a deal—it is a virtual entity that may (but doesn't necessarily have to) lead to physical product. Common synonyms for eBusiness include e-Commerce and e-Trading.
EDI VAN	A value added network that offers protocol and line speed matching to facilitate communication between trading partners and other EDI-related services such as carbon copy service, conversion of EDI transaction sets to fax or mail, and trading partner implementation services. See also Electronic Data Interchange and Value Added. Also referred to as a clearinghouse.
EDIFACT	EDI for Administration, Commerce, and Transport. This standard contains data requirements for carrying on international trade, and has been accepted by countries all over the world as their EDI standard.

Electronic Commerce	The sale or procurement of supplies and services using information systems technology.
Electronic Data Interchange	(1) Intercompany, computer-to-computer communication of data which permits the receiver to perform the function of a standard business transaction and is in a predefined standard data format. (2) The transfer of structured data, using agreed-upon message standards, from one computer system to another, by electronic means.
Electronic Funds Transfer	Computerised systems that process financial transactions and information about financial transactions; specifically, the exchange of value between two financial institutions.
Email	Common abbreviation for electronic mail. Email comes in many guises and has been popularised, to a large extent, through the growth of the Internet.
Encryption	Encryption is the process of disguising a message (using mathematical formulas called algorithms) in such a way as to hide its substance, a process of creating secret writing.
Enterprise network	The collection of public and private switches, transmission links, sub-networks, management systems, network applications, etc. that combine to provide an identifiable group of people with the communication service that they need to operate effectively.
Fault tolerance	A method of ensuring that a computer system or network is more resilient to faults or breakdowns, to avoid lost data and downtime. Differing applications achieve this, and include processor duplication and redundant media systems.
FAQ	Frequently Asked Question. A set of files, available over the Internet, that provide a compendium of accumulated knowledge in a particular subject.
File server	A station in local area networks dedicated to providing file and data storage to other terminals in the network.
FTP	File Transfer Protocol. The Internet standard high-level protocol for transferring files from one computer to another. A widely used *de facto* standard (cf. the sparingly used *de jure* standard above).
FYI	For Your Information. These are Internet bulletins that answer common questions.
Gateway	Hardware and software that connect incompatible networks, which enables data to be passed from one network to another. The gateway performs the necessary protocol conversions.
GOS	Grade Of Service. The probability that a call or connec-

	tion will be blocked. This relates to circuit switched networks, and is determined by the amount of switching and transmission equipment provided (per user) in the network.
GOSIP	Government Open Systems Interconnect Profiles. A UK initiative to help users procure open systems.
GUI	Graphics User Interface—an interface that enables the user to select a menu item by using a mouse to point to a graphic icon (small simple pictorial representation of a function such as a paint brush for shading diagrams, etc.). This is an alternative to more traditional character-based interface, where an alphanumeric keyboard is used to convey instructions.
Hardware	The physical equipment in a computer system. It is usually contrasted with software.
Hierarchical network	A network structure composed of layers. An example of this can be found in a telephone network. The lower layer is the local network followed by a trunk (long-distance) network up to the international exchange networks.
Host	Commonly used synonym for server.
Hostage data	Data which is generally useful but held by a system which makes external access to the data difficult or expensive.
HTML	Hyper Text Markup Language is used to describe formatting in web documents. It is defined in the Internet's RFC series.
HTTP	Hyper Text Transfer Protocol. This is the basic protocol that underlies the World-Wide Web. It is a simple, stateless request-response protocol, defined in the Internet's RFC series.
IAB	Internet Activities Board. The influential panel that guides the technical standards adopted over the Internet. Responsible for the widely accepted TCP/IP family of protocols. More recently, the IAB have accepted SNMP as their approved network management protocol.
Indirect	Purchases which do not go directly into the end-product of the buying organisation. Account for over 80% of transaction volume. Generally not well-automated and are generally subject to sales taxes.
Information processor	A computer-based processor for data storage and/or manipulation services for the end user.
Information retrieval	Any method or procedure which is used for the recovering of information or data which has been stored in an electronic medium.

Information Technology	A rather loosely defined term that is usually taken to cover the technology relevant to providing information services for an organisations. Generally a mix of software, hardware, office systems databases, networks and computing.
Integrated Financial System	A software application which interfaces with the finance, accounting and general ledger functions.
Interface	The boundary between two things, typically two programs, two pieces of hardware, a computer and its user, a project manager and the customer.
Internetworking	Specialised form of interworking where the interaction involves two or more networks.
Interworking	Generic interaction between two entities to achieve operational, syntactic and semantic integrity of information.
Inventory	Wherever goods are stored and materials management is applied.
Invoice	A document prepared by the seller to apprise the buyer of the charge for product purchased and shipped by the seller.
ISDN	Integrated Services Digital Network—an emerging end-to-end CCITT standards for voice, data and image services. The intention is for ISDN to provide simultaneous handling of digitised voice and data traffic on the same links and the same exchanges.
Just-In-Time Delivery	Tight integration of the supply chain to allow buying organisations to achieve drastic reduction o internal stocks.
Key	The record identifier used in many information retrieval systems (i.e. database keys).
LAN	Local Area Network—a data communications network used to interconnect data terminal equipment distributed over a limited area.
Language	An agreed-upon set of symbols, rules for combining them and meanings attached to the symbols that is used to express something (e.g. the Pascal programming language, job-control language for an operating system and a graphical language for building models of a proposed network).
LDAP	Local Directory Access Protocol. A protocol used on IP-based networks for populating customer records.
Legacy system	A system which has been developed to satisfy a specific requirement and is, usually, difficult to substantially re-configure without major re-engineering. Sometimes referred as a cherished system (and sometimes as a millstone).

Lifecycle	A defined set of stages through which a piece of software passes over time—from requirements analysis to maintenance. Common examples are the waterfall (for sequential, staged developments) and the spiral (for iterative, incremental developments). Lifecycles do not map to reality too closely, but do provide some basis for measurement and hence control.
Line Item	A unique item from a product line on which inventory and other records are kept. Usually identified with a specific number for reference.
Line	An individual line item entry on an order or order request.
Logistics	All activities associated with transporting and storing goods and services.
Maintenance	Changes to software or a network after its initial development; also called evolution. In practice, it is the task of modifying (locating problems, correcting or updating, etc.) hardware configuration and (more often) software systems after they have been put into operation.
Matching	An accounting concept developed to prevent fraud and errors, the most complete being a 'Three Way Match' reconciling the Purchase Order, Receiving Record and Invoice.
Materials Management	A scientific methodology for planning, managing, and tracking inventory and information.
Materials Requirements Planning	A subset of the planning aspect of Materials Management.
MD5	A secure hashing algorithm that converts an arbitrarily long data stream into a digest of fixed size. This algorithm is widely used for insuring message integrity and as part of creating and verifying digital signatures.
Merchant Server	A server configured for electronic commerce.
Method	A way of doing something. It is generally a defined approach to achieving the various phases of the lifecycle. Methods are usually regarded as functionally similar to tools (e.g. a specific tool will support a particular method).
Middleware	A general terms for software that provides connectivity between front-end and back-end systems. The latter are usually legacy systems that hold important data. The middleware provides a common set of services, interfaces and functions that can be used by new front-ends. The technology that enables middleware is DCE, DCOM and remote procedure calls.

Model	An abstraction of reality that bears enough resemblance to the object of the model that we can answer some questions about the object by consulting the model.
Modelling	Simulation of a system by manipulating a number of interactive variables; can answer 'what if. . .?' questions to predict the behaviour of the modelled system. A model of a system or sub-system is often called a prototype.
Modularisation	The splitting up of a software system into a number of sections (modules) to ease design, coding, etc. Only works if the interfaces between the modules are clearly and accurately specified.
MRO	Maintenance and Repair Organisation. Term also commonly used to refer to low-dollar, high volume transactions.
NACHA	The National Automated Clearing House Association. The banking trade association that is responsible for developing and maintaining the rules for the ACH Network. More than 20,000 financial institutions, 400,000 companies and businesses, and millions of consumers use and benefit from the ACH payment system.
Network	A general term used to describe the inter-connection of computers and their peripheral devices by communications channels. For example, Public Switched Telephone Network (PSTN), Packet Switched Data Network (PSDN), Local Area Network (LAN), Wide Area Network (WAN).
Network interface	The circuitry that connects a node to the network, usually in the form of a card fitted into one of the expansion slots in the back of the machine. It works with the network software and operating system to transmit and receive messages on the network.
Network management	A general term embracing all the functions and processes involved in managing a network, and include configuration, fault diagnosis and correction. It also concerns itself with statistics gathering on network usage.
Network topology	The geometry of the network relating to the way the nodes are interconnected.
NFS	Network file system. A method, developed by Sun microsystems, that allows computers to share files across a network as if they were local.
Non-proprietary	Software and hardware that is not bound to one manufacturer's platform. Equipment that is designed to the specification that can accommodate other companies' products. The advantage of non-proprietary equipment

	is that a user has more freedom of choice and a larger scope. The disadvantage is when it does not work, you may be on your own.
NIC	The Network Information Centre. Source of much information on the Internet and related networking issues.
OBI EDI convention	This is the EDI-based specification for the format of an OBI order request or order which is based on the EDI ASC X.12 850 standard for electronic purchase orders.
OBI order request	This is an EDI-based data structure that reflects the contents of a requisitioner's 'shopping cart'. It is sent (within an OBI object) from a Selling Organization to the requisitioner's organisation for order completion and approval. The format of an order request is specified by the OBI order format convention which is based on the ASC X.12 850 EDI standard.
OBI order	This is an EDI-based data structure that reflects an official, authorized request for goods and services based on pre-defined pricing, terms and conditions. It is sent (within an OBI object) from a buying organisation to a trading partner and is typically associated with a related order request. The format of an order is specified by the OBI order format convention which is based on the ASC X.12 850 EDI standard.
OBI	Open Buying on the Internet. A standard for secure, interoperable, business-to-business Internet commerce developed by the Internet Purchasing Roundtable and maintained by the OBI Consortium.
OBI WAN	True first name of Ben Kenobe, Jedi knight.
One-way hash function	A one-way transformation that converts an arbitrary amount of data into a fixed-length hash. It is computationally hard to reverse the transformation or to find collisions. MD5 and SHA are examples of one-way hash functions. One-way hash functions are utilised within digital signatures.
Object	An abstract, encapsulated entity which provides a well defined service via a well defined interface.
Object oriented	A philosophy that breaks a problem into a number of co-operating objects. Each object has defined properties (e.g. it can inherit features from another object). Object-oriented design is becoming increasingly popular (e.g. in the specification of network management systems).
Object program	The translated versions of a program that has been assembled or compiled.

OMG	Object Management Group. The OMG is responsible for the CORBA initiative.
Order Shipping Set	Generally for direct purchases. Includes order picking instructions, bill of lading, stock status update, and invoice information.
Order Status	Where an order stands in the process.
Packet switching	The mode of operation in a data communications network whereby messages to be transmitted are first transformed into a number of smaller self-contained message units known as packets. Packets are stored at intermediate network nodes (packet-switched exchanges) and are reassembled unto a complete message at the destination. A CCITT recommendation standard for packet switching is X.25.
Parameter	A variable whose value may change the operation but not the structure of some activity (e.g. an important parameter in the productivity of a program is the language used).
Peer to peer	Communications between two devices on an equal footing, as opposed to host/terminal, or master/slave. In peer-to-peer communications both machines have and use processing power.
Pixel	Picture element—the smallest discrete element making up a visual display image.
PKCS #7	Cryptographic Message Syntax Standard from RSA Data Security. Defines a general syntax for data with cryptography applied to it. OBI uses the PKCS #7 standard for digital signatures.
PKCS	Public Key Cryptography Standards. A set of cryptographic standards developed by RSA Data Security.
Polling	Process of interrogating terminals in a multipoint network in turn in a prearranged sequence by controlling the computer to determine whether the terminals are ready to transmit or receive. If so, the polling sequence is temporarily interrupted while the terminal transmits or receives.
Port	A device which acts as an input/output connection. Serial port or parallel port are examples. Also, a pleasing after dinner drink.
Preferred Selling Organisation	A selling organisation with an existing contractual agreement with a buying organisation which meets pre-specified performance criteria, and through whom the buying organisation attempts to channel buying for a pre-defined set of products and services.

Presumptive Receipt	Less formal than Evaluated Receipt, delivery of goods and services is presumed if an Invoice is received.
Private Key Encryption	In this scheme, the security of the encryption depends on a shared secret that only the two communicating parties know. The International Data Encryption Algorithm (IDEA) and Data Encryption Standard (DES) are examples of private key systems.
Process	Technically, a procedure that is being executed on a specific set of data; more generally, a procedure for doing something that is actually being carried out.
Program	A set of instructions for a computer, arranged to that when executed they will cause some desired effect (such as the calculation of a quantity or the retrieval of a piece of data).
Programming language	An artificial language constructed in such a way that people and programmable machines can communicate with each other in a precise and intelligible way. Fortran, Cobol and C are three languages that account for most of the deployed software systems at present.
Proprietary	Any item of technology that is designed to work with only one manufacturer's equipment. The opposite of the principle behind Open Systems Interconnection (OSI).
Protocol	A set of rules and procedures that are used to formulate standards for information transfer between devices.
PSTN	Public Switched Telephone Network—the public telephone system providing local, long-distance and international telephone services. Widely used (with modems) for many other data services.
Public Key Encryption	In this scheme, a user has a pair of keys—one private and one public. A message encrypted using the public key can only be decrypted using the private key and vice versa. So you can receive messages from anyone who knows your public key (which you decrypt with your private key) and can happily send an encrypted message to anyone who's public key you know. Perhaps the best known of the public key systems is RSA.
Purchase Order	A document issued by a buyer to a seller that details the terms of sale under which the buyer will purchase the seller's goods.
PKCS	Public-Key Cryptography Standards. A set of standards for public-key cryptography, developed by RSA Laboratories in cooperation with an informal consortium, originally including Apple, Microsoft, DEC, Lotus, Sun and MIT.

	Published standards are PKCS #1, #3, #5, #6, #7, #8, #9, #10 and #11. PKCS includes both algorithm-specific and algorithm-independent implementation standards. Algorithms supported include RSA and Diffie–Hellman key exchange among many others. However, only RSA and Diffie–Hellman are specifically detailed. It also defines an algorithm-independent syntax for digital signatures, digital envelopes, and extended certificates; this enables someone implementing any cryptographic algorithm whatsoever to conform to a standard syntax, and thus achieve interoperability.
Purchase Order	A document directing a seller to supply goods or services.
Purchase Price Variance	The difference between the expected cost of an order and the invoiced cost of the order. Is sometimes measured on a line item basis.
Purchasing Card	A card based system, issued by a Payment Authority or other financial institution, which is used to purchase low dollar goods and services.
Purchasing	The function or department (within an organisation) responsible for buying/obtaining goods and services. See buying organisation.
Queuing	When a frame or packet is to be transmitted on a link, it may have to wait because another frame is being processed in front of it. The frame is placed in a buffer until the transmitter is free. Hence, queuing systems (i.e. packet switched systems) require buffers (matched to load and capacity) and introduce delay (as opposed to circuit switching systems, which block).
Resolve	Translate an Internet name into it's equivalent IP address or other DNS information.
RFC	Request for Comment. A long-established series of Internet informal documents widely followed by commercial software developers. RFCs tend to provide the implementation detail to supplement the more general guidance of ISO and other formal standards. The main vehicle for the publication of Internet standards, such as SNMP.
RSA	A well known and widely used software-based public key encryption method. It is named after its inventors—Rivest, Shamir and Adleman.
SCSI	Small Computer System Interface. A bus-independent standard for system level interfacing between a computer and an intelligent device (e.g. an external disk). Pronounced 'scuzzy'.

SSL Secure Sockets Layer. The Secure Sockets Layer Handshake Protocol was developed by Netscape Communications Corporation to provide security and privacy over the Internet. The protocol supports server and client authentication. The SSL protocol is application independent, allowing protocols like HTTP, FTP (File Transfer Protocol), and Telnet to be layered on top of it transparently.

The SSL protocol is able to negotiate encryption keys as well as authenticate the server before data is exchanged by the higher-level application. The SSL protocol maintains the security and integrity of the transmission channel by using encryption, authentication and message authentication codes.

Secure When the level of security is commensurate with the business risk.

Security The combination of software, hardware, networks and policies designed to protect sensitive business information and to prevent fraud.

Server An object which is participating in an interaction with another object, and is taking the role of providing the required service.

Session The connection of two nodes on a network for the exchange of data—any live link between any two data devices.

Settlement In banking, settlement refers to the process of recording the debit and credit positions of two parties in a transfer of funds. Also, it is the delivery of securities by a seller and the payment by the buyer.

Settlement banks These are banks that maintain the settlement accounts for clearing members whereby payments and deposits are made.

SHA The secure hash algorithm defined in FIPS PUB 180-1. It produces a 20-byte output.

Shipment An individual release of product to a specific requisitioner. A shipment may include the entire order, a part of an order, or a consolidation of multiple orders.

Signalling The passing of information and instructions from one point to another for the setting up or supervision of a telephone call or message transmission.

SMTP Simple Mail Transfer Protocol. The Internet standard for the transfer of mail messages from one processor to another. The protocol details the format and control of messages.

Specification	A description of a system or program that states what should be provided but does not necessarily provide information on exactly how the system or program will work.
SQL	Structured Query Language. Probably the most widely used database access language. SQL comes in a variety of flavours.
Stock Keeping Unit	A number generated by a customer organisation to uniquely identify a product. The Universal Product Code (UPC), a universally (by vendor and customer) accepted number, may be used as an SKU number.
Stockout	An inventory condition where a specific line item is not available for prompt shipment.
Store and Forward	Pertaining to transmission of data, messages are received at an intermediate routing point and recorded (stored). They are subsequently transmitted to a further routing point or to the ultimate recipient.
Supply Chain Management	The strategies and activities associated with the supply chain.
Supply Chain	All organisations and processes related to products and services sourced by buying organisations. Typically covers everything from raw materials through consumption.
SWIFT	The Society for Worldwide Interbank Financial Telecommunication is an international financial payment cooperative organisation that operates a network that facilitates the exchange of payment and other financial messages between financial institutions throughout the world.
Syntax	The set of rules for combining the elements of a language (e.g. words) into permitted constructions (e.g. phrases and sentences). The set of rules does not define meaning, nor does it depend upon the use made of the final construction.
System	A collection of independently useful objects which happen to have been developed at the same time.
Systemic risk	This refers to the risk that the failure of one participant in a transfer system (or financial markets generally) to meet its required obligations will cause other participants or financial institutions to be unable to meet their obligations when due.
TCP/IP	Transmission Control Protocol/Internet Protocol—a set of transport internetworking protocols that have become a *de facto* networking standard. Commonly used over X.25 and Ethernet wiring, they are viewed as two of the few protocols available, that are able to offer a true

	migration path towards OSI. TCP/IP operates at the third and fourth layers of the OSI model (network and transport, respectively).
Telecommunications	The general name given to the means of communication information over a distance by electrical and electromagnetic methods. The transmission and reception of information by any kind of electromagnetic system.
Telnet	A TCP/IP based application that allows connection to a remote computer.
Third Party Purchases	Spot buys sourced through a preferred selling organisation on behalf of a buying organisation.
Throughput	A way of measuring the speed at which a system, computer or link can accept, handle and output information.
Topology	A description of the shape of a network, for example star, bus and ring. It can also be a template or pattern for the possible logical connections onto a network.
TP	Transaction Processing. Concerned with controlling the rate of enquiries to a database. Specialist software—known as a TP monitor—allows a potential bottleneck to be managed.
Trade comparison	This is the receipt, validation and matching of data on the long (buy) and short (sell) side of a transaction and the reporting of such match.
Trading Partner	One of at least two participants in an established business relationship.
Transaction Set	A collection of formatted data that contain the information required by a receiver to perform a standard business transaction. In an EDI standard, a transaction set is defined as having three sections, header, detail and summary, and is comprised of a predefined group of segments in each section.
Transaction	Exchange of business information pertinent to an agreement or deal.
Transfer encoding	This is a reversible conversion of a data stream prior to transmission to ensure proper processing. Typically this is done so that 8 bit data may be sent via a channel that only transmits 7 bit data.
UCC	The Uniform Commercial Code (UCC) is a set of model laws governing commercial and financial transactions.
UMTS	Universal Mobile Telephony Service. This is the concept that users will be able to access all of their network services irrespective of location. The goal of UMTS is to provide a network based around the individual—the individual registers their access device with the network rather than having to find a fixed access device.

UN/EDIFACT	Electronic Data Interchange For Administration, Commerce and Transportation. EDIFACT is a set of message standards approved by the United Nations for international electronic data interchange.
Verification	This is the ability to positively identify and authenticate a particular encrypted communication.
Value Added Network	A company that acts as a pipe or an electronic mailbox for the transmission of data, and provides communications services such as line speed conversion and protocol matching.
Vendor	One that offers wares for sale, a selling organisation.
Vendor independent	Hardware or software that will work with hardware and software manufactured by different vendors—the approximate opposite of proprietary.
Virtual circuit	A logical connection across a network (e.g. the transmission path through an X.25 packet-switched data network established by exchange of set up messages between source and destination terminals).
Virtual device	A module allocated to an application by an operating system or network operating system, instead of a real or physical one. The user can then use a computer facility (keyboard or memory or disk or port) as though it was really present. In fact, only the operating system has access to the real device.
Virus	A program, passed over networks, that has potentially destructive effects once it arrives. Packages such as Virus Guard are in common use to prevent infection from hostile visitors.
VPN	Virtual Private Network. A combination of public and private resources that has been combined to give the user a network that looks like a coherent resource, suited to their particular needs. To all intents and purposes, a VPN is an Enterprise Network.
Web Catalogue	An ordered set of vendor offerings available via the World-Wide Web.
Wide Area Network	A connection of the computers of several business users within an organisation such that they may all access the same computer software or may have access to one another's computer files.
Workflow	The process of collaborative work within an organisation.
Worldwide Web	An Internet-based project that has provided an intuitive and powerful means of navigating a very large information space. The Web has become almost synonymous with the client packages (browsers) such as Netscape and Internet Explorer that allow a user to traverse the

	Internet via hypertext links to access multimedia information.
Worm	A computer program which replicates itself—a form of virus. The Internet worm was probably the most famous—it successfully, and accidentally, duplicated itself across the entire system.
X12	The ANSI committee responsible for the development and maintenance of standards for EDI.
X509.V3	ITU-T Recommendation X.509 specifies the widely adopted X.509 public key certificate syntax. A certificate is signed by the certificate issuer to authenticate the binding between a specific user's name and public key.
X/Open	An industry standards consortium that develops detailed system specifications drawing on available standards. X/Open owns the UNIX trademark and thereby brings focus to its various flavours (e.g. HP-UX, AIX from IBM, Solaris from SUN, etc.).
XML	eXtended Markup Language. A more powerful, and general, successor to HTML that allows the nature of information embedded in a web page to be identified as well as its presentation format.

BIBLIOGRAPHY

Discovery consists of seeing what everyone has seen and thinking what nobody has thought

Nagyrapolt Szent-Gyorgi

In many ways, there is nothing new about electronic business; it is simply a way of conducting business using the latest technology. But as the above quote suggests, innovation often comes about when different schools of thought are focused on a common goal. So it is in this instance. The science that makes eBusiness a reality draws on distributed computing, software engineering, network and information technology, to name but a few.

There may not be many books on eBusiness *per se* (at least not that go into a lot of depth) but there are plenty of excellent texts that cover part of its practical foundations. Here are a few of our favourites.

Online business

Cronin M, *Doing Business on the Internet*, Van Nostrand Reinhold, 1994.
Ellsworth J, *The Internet Business Book*, John Wiley & Sons, 1994.
Lynch D & Lundquist L, *Digital Money*, John Wiley & Sons, 1995.
Yesil M, *Setting Up the On-line Shop*, John Wiley & Sons, 1998.

Security

Oppliger R, *Internet and Intranet Security*, Artech House, 1998.
Cheswick W & Bellovin S, *Firewalls and Internet Security: Repelling the wily hacker*, Addison-Wesley Longman, 1994.

Networked systems

Atkins J & Norris M, *Total Area Networking*, John Wiley & Sons, 1998.
Coulouris G, Dollimore J & Kindberg T, *Distributed Systems—Concept and Design*, Addison-Wesley, 1994.
Subramanian M, *Introduction to Network Management*, Addison-Wesley Longman, 1999.

The information business

Whittaker *et al.*, *Strategic Systems Planning*, John Wiley & Sons, 19??.
Cooke P, Moulaert F, Swyngedouw E, Weinstein O & Wells P, *Towards Local Globalisation*, UCL Press, 1993.
Naisbitt J, *Global Paradox*, Nicholas Brealey Publishing, 1994.
Handy C, *The Age of Unreason*, Arrow, London, 1990.
Ray P, *Computer Supported Cooperative Work (CSCW)*, Prentice Hall, 1999.

Component technologies

Thomas A, *Enterprise JavaBeans: Server Component Model for Java*, Patricia Seybold Group, December 1997.
Norris M, Davis R & Pengelly A, *Component Based Network Systems Engineering: Interfaces and Integration*, Artech House, 1999.

Software architecture

Bass L, Clements P & Kazman R, *Software Architecture in Practice*, Addison-Wesley, 1998.
Objects, Methods & Architecture, *IEEE Software*, Special Edition, January/February 1997.
Gamma E, Helm R, Johnson R & Vlissides J, *Design Patterns: Elements of Reusable Object-Oriented Software*, Addison-Welsey, 1994.

Index